Sustainable Development Goals Series

The **Sustainable Development Goals Series** is Springer Nature's inaugural cross-imprint book series that addresses and supports the United Nations' seventeen Sustainable Development Goals. The series fosters comprehensive research focused on these global targets and endeavours to address some of society's greatest grand challenges. The SDGs are inherently multidisciplinary, and they bring people working across different fields together and working towards a common goal. In this spirit, the Sustainable Development Goals series is the first at Springer Nature to publish books under both the Springer and Palgrave Macmillan imprints, bringing the strengths of our imprints together.

The **Sustainable Development Goals Series** is organized into eighteen subseries: one subseries based around each of the seventeen respective Sustainable Development Goals, and an eighteenth subseries, "Connecting the Goals", which serves as a home for volumes addressing multiple goals or studying the SDGs as a whole. Each subseries is guided by an expert Subseries Advisor with years or decades of experience studying and addressing core components of their respective Goal.

The SDG Series has a remit as broad as the SDGs themselves, and contributions are welcome from scientists, academics, policymakers, and researchers working in fields related to any of the seventeen goals. If you are interested in contributing a monograph or curated volume to the series, please contact the Publishers: Zachary Romano [Springer; zachary.romano@springer.com] and Rachael Ballard [Palgrave Macmillan; rachael.ballard@palgrave.com].

Tinuade Adekunbi Ojo
Editor

Women and Finance in Africa

Inclusion and Transformation

Springer

Editor
Tinuade Adekunbi Ojo
Department of Politics and International Relations
University of Johannesburg
Johannesburg, South Africa

ISSN 2523-3084 ISSN 2523-3092 (electronic)
Sustainable Development Goals Series
ISBN 978-3-031-53336-5 ISBN 978-3-031-53337-2 (eBook)
https://doi.org/10.1007/978-3-031-53337-2

This Springer imprint is published by the registered company Springer Nature Switzerland AG
The registered company address is: Gewerbestrasse 11, 6330 Cham, Switzerland

Paper in this product is recyclable.

Preface

Africa is the leading region in the world in the expansion of mobile money transactions, according to Global Findex. The purpose of this book is to present the different financial inclusion policy approaches in African countries and their preparedness for the technologies that enable financial inclusion for women across all disciplines. This book intends to analyse Fintech Initiatives in Africa, how Fintech has Fared in Africa and areas of Africa's cooperation in growing digital financial inclusion. The findings intend to analyse the effectiveness of the countries' approaches and share lessons that different African economies, whether currently booming or struggling, can enhance or implement towards financial inclusion at all structural levels.

Despite the importance of financial inclusion in response to the growth and development of the economy, critics have argued that financial inclusion represents regressive policies that have hindered the government from meeting the targeted ideological goals set for each country. The hindrance might be traced to the 2007–2008 global financial crisis, which crashed the global economy. Many countries, especially developing economies, are yet to fully recover and achieve the set goals on financial inclusion for their economies (Prabhakar, 2019: 40). This is reiterated by Langley (2008) and Leyshon et al. (2008: 6), who argue that most developing economies have abandoned financial inclusion and the government has passed the responsibilities to its people. The themes identified will provide guidance on the compilation of state-specific profiles on different national approaches to financial inclusion gender policies. The main objective of this volume is to understand different processes for financial inclusion to gender issues at a national level. And to help encourage reflection on what lessons could be learnt between states and what factors cause divergence in multilateral settings so that they can be understood and hopefully addressed.

Johannesburg, South Africa

Tinuade Adekunbi Ojo

Synopsis

The subject of financial inclusion and women empowerment lies at the heart of discussions about inequality and how it may be remedied (Kara et al., 2021). These discussions have become prominent these days especially through the facilitation of UN Women for instance and other structures seeking to enable women's empowerment (Cicchiello et al., 2021). The discussion is fast moving from nominal inclusion of women in existing political and economic structures towards enabling women's own initiatives and agency to influence outcomes of gender transformation of politics and economy. According to the UN Capital Development Fund, financial inclusion is positioned prominently as an enabler of other developmental goals in the 2030 Sustainable Development Goals, where it is featured as a target in 8 of the 17 goals (Pauliukevičienė & Stankevičienė, 2022). These include SDG1, on eradicating poverty; SDG 2 on ending hunger, achieving food security, and promoting sustainable agriculture; SDG 3 on profiting health and well-being; SDG 5 on achieving gender equality and economic empowerment of women; SDG 8 on promoting economic growth and jobs; SDG 9 on supporting industry, innovation, and infrastructure; and SDG 10 on reducing inequality (Smith, 2022). Additionally, in SDG 17 on strengthening the means of implementation there is an implicit role for greater financial inclusion through greater savings mobilization for investment and consumption that can spur growth (Smith, 2022). This emphasizes the role of financial inclusion in the current global discourse on gender equality, women empowerment, and the quest to us institutions such as financial institutions for resource allocation and poverty alleviation or socio-economic sustainability in the broader perspective.

This book seeks to contribute to shifting the geography of thinking about women's financial inclusion from broad discussions to specific vantage points held by women. It is thus designed to enable society to learn from women actors on the ground. This book should therefore be helpful to thinkers who are thinking about new directions in women empowerment, researchers who are finding ways of overcoming impediments to women's financial inclusion, practitioners working practically to unblock stumbling blocks in financing institutions, and readers hoping to learn about how to understand factors producing exclusion or enabling inclusion for women's empowerment.

The book aims to analyse conditions explaining financial inclusion and women's empowerment in Africa. Throughout the book, the authors explore the issues on the state of financial inclusion and exclusion of women in Africa, the relationship between financial inclusion and socio-economic development, the existing government measures on financial inclusion, and whether they contribute to women's economic empowerment. The book also debates the socio-economic factors enabling and limiting women's access to and using financial products to improve their socio-economic empowerment and the future suggestions, policies, and recommendations on financial inclusion for women in Africa. In addressing these questions, the authors argue that there had been inadequate financial inclusion/exclusion analysis, especially concerning women in Africa. Among the gaps identified on financial exclusion are the challenges encountered in the current model of financial services: the fact that most women remain restricted to the household's financial responsibilities and have limited exposure to the benefits of financial institutions and mechanisms that can guarantee financial freedom help to save money and eradicate poverty (WEF, 2016; Klapper et al., 2017; World Bank, 2017).

This book presents the different financial inclusion policy approaches in African countries and their preparedness for the technologies that enable financial inclusion for women across all disciplines. First, this book intends to analyse Fintech Initiatives in Africa, how Fintech has Fared in Africa and areas of Africa Cooperation in growing digital financial inclusion. Second, the book presents several significant themes and African states' efforts to address the political and economic factors influencing budget allocation to women-oriented programmes and projects in African communities. The aim is to investigate the notion of inclusion and transformation in African economies. The book's outcome intends to analyse the effectiveness of the countries' approaches and share lessons from different African economies on financial inclusion for women. Contributions to the book will adopt a gender policy-centric approach specifying the role of "Sustainable Development Goals" on each theme to be contributed.

Structure of the Book

Women Finance in Africa: Inclusion and Transformation presents the assumptions, narratives, and institutions that underpin the key concepts and investigate the limits and potential of financial inclusion development strategy for gender equality. Despite the importance of financial inclusion in response to the growth and development of the economy; critics have argued that financial inclusion represents regressive policies that have hindered the government from meeting the targeted ideological goals set for each country. The hindrance might be traced to the 2007–2008 global financial crisis, which crashed the global economy. Many countries, especially developing economies, are yet to fully recover and achieve the set goals on financial inclusion for their economies (Prabhakar, 2019: 40). This is reiterated by Langley (2008) and Leyshon et al. (2008: 6), who argue that most developing economies have abandoned financial inclusion and the government has passed the responsibilities to its people. The themes identified will provide guidance on the compilation of state-specific profiles on different national approaches to financial inclusion gender policies. The main objective of this volume is to understand different processes for financial inclusion to gender issues at a national level and to help encourage reflection on what lessons could be learnt between states and what factors cause divergence in multilateral settings so that they can be understood and hopefully addressed.

The book begins with Tinuade Ojo's useful overview of financial inclusion and structure of the book. The chapter explores the complexities surrounding gendered finance in Africa, emphasising the pivotal role of inclusion and transformation in reshaping financial structures. The chapter further underscores the imperative of addressing gendered financial inequalities in Africa, emphasising the need for holistic approaches that combine policy reforms, innovative financial products, and cultural shifts. By embracing inclusion and transformation in financial systems, the author concludes that, Africa can unlock the immense potential embedded in its diverse population, fostering a more equitable and prosperous future for all. Chapter 2, on the other hand, explores Eric Magale and Edson Chiwenga's manifestations on the informal financial services in the financial inclusion matrix: gendered perspectives from Kenya and Zimbabwe. The chapter assesses the confluence of informal financial services in Africa and the global discourse on financial inclusion with particular reference to women. It further argues that informal financial services continue to play a salient role among women even in the face of rapid adoption of financial technology and a fast-developing financial industry in the continent. The chapter also argues for a broader understanding of financial inclusion with reference to women in Africa and the use of informal financial services. Chapter 3 unravels Queensley Chukwudum's debate on the understanding of financial inclusion in Africa from the perspective of saving and burrowing patterns. The chapter examines the savings and burrowing patterns of Africans through the gender and income status lenses. The paper composition indicates a strong inclination towards saving with a savings club for women compared to saving with a financial institution. Chapter 4 narrates Michelle Beeslar's position on "the political socioeconomic fallout of the mismanagement of COVID-19 public funds in South Africa: gendered finance for women in agribusiness. According to the author, the inability of women to access their constitutional right to sufficient food and water has enhanced limited opportunities for education, employment and entrepreneurship. The global political and socioeconomic impact of the COVID-19 pandemic on women was worsened by the mismanagement of social response public funds; the study posits that there is an adequate state policy framework coupled with enough public sector resources, both budgetary and human to facilitate food security and empowerment for women.

In Chap. 5, Purity Hamunakwadi narrates the theme on re-imagining South African women informal entrepreneurship as spaces for sustainable local economic development praxis: a financial inclusion perspective. The chapter attempts to interrogate and shed light on the role that financial inclusion can play towards supporting women informal entrepreneurship taking lessons from other similar experiences across the world. Chapter 6 unravels Lena Magerere and Edwin Mtengwa's study on the funding challenges women entrepreneurs face in the agricultural sector in Zimbabwe and the causes thereof. According to them, access to funding is cited by many scholars as one of the biggest challenges faced by entrepreneurs in Africa. Zimbabwean entrepreneurs are not spared from this problem. Business funding for entrepreneurs in Zimbabwe has been problematic across the entire business fraternity since immemorial. This scenario is more prevalent in the agricultural sector with women entrepreneurs bearing thc most brunt.

Chapter 7 explains Belinda Chaora's position on "The Role of Institutions in Promoting Financial Inclusion of MSMEs in Zimbabwe". This chapter processes data from a nationwide survey of MSMEs as primary data and subsequently uses secondary sources to compile other informal institutions influencing the financial habits of MSMEs in Zimbabwe. The chapter outlines some of the barriers experienced by MSMEs when attempting to access products and services form formal sources. It further outlines the key players in financial service provision and their roles noting that other institutions (social and cultural) should additionally be considered in future efforts to understand the state of inclusion and promote financial equity in Zimbabwe.

Chapter 8 narrates Temitayo Adeyemo's notion on the financial inclusion for women in Nigeria: much ado about access. The chapter examines financial inclusion from the multidimensional lens of opportunity, accessibility, and utilization for Nigerian women in general and rural women in particular. Findings reveal that there are demand and supply factors that lead to self-selection of women into degrees within the financial inclusion frame, with differentials across rural and urban sectors. Chapter 9 presents Omosefe Oyekanmi and Agunbiade Doyinsola notion for a broader understanding of financial inclusion concerning women in Africa and the use of informal financial services. The chapter examines the financing of women and their inclusion in the Ivorian peacebuilding process. It investigated women's role during the Ivorian civil wars and their contribution to building peace. Using a case study research design, secondary data were sourced and content was analysed through the qualitative method. The study revealed that from victims of sexual abuse to forceful conscription as combatants, the role of women during the civil wars remains complex. More so, through formal and informal structures, women were essential to the nation's efforts during the post-conflict reconstruction period. In Chap. 10, Rifkatu Nghargbu and Fadila Jumare on the title "Digital Financial Inclusion for Women in Africa: Prospects and Challenges" discusses how women digital financial inclusion can be achieved in Africa, its prospects for women empowerment and likely challenges. The book chapter argues that the growth in digital finance can be attributed to increase in the ownership of mobile phones which has triggered tremendous increase in innovative digital tools, transactions, and services worldwide. However, in many African countries most women don't have access to digital financial services, especially those living in poor and marginalized communities, due to their inability to own digital assets such as smart phones, internet services, and the knowledge to perform digital transactions. Chapter 11 highlights Kemi Akeju's position on Gender and Digital Financial Inclusion: Assessment of Rural/Urban Dichotomy in Microfinance and Table Banking Operations in ECOWAS. This chapter using selected ECOWAS countries' experiences assesses the challenges, benefits, and risks faced by the traditional microfinance institutions and operators of table banking platforms in their digitalization process towards enhancing gender equality in financial services. It considers ways of bridging the gender divide in assessing financial services through the partnership of microfinance and table banking operators in rural communities with digital financial services operators given the rise in recognition of the need to include women and the large unbanked society in rural communities. Furthermore, it addresses factors that can advance women's access to digital financial services, build trust in its effectiveness, and promote equality of opportunities in rural areas. Chapter 12, presents Blessing Olanrewaju and Yetunde Aluko's position on Gendered Dimension for Digital Financial Inclusion in Nigeria. It highlights the gender gap in access to financial services and the role of digital finance in addressing this gap in Nigeria. The findings confirmed the pervasiveness of gender gaps in access to digital financial services in Nigeria as more men than women have access to digital finance platforms.

The book uses a qualitative research method and presents both primary to secondary data to narrate the impact of gender-responsive budgeting on women's empowerment and gender equality in these communities. The chapters present the analysis of the effectiveness of African state's approaches and share lessons that different African economies, whether currently booming or struggling, can enhance or implement towards the financial inclusion and gender budgeting response at all structural levels.

Abbreviations

ACHPR	African Charter on Human and People's Rights
AFAWA	Affirmative Financial Action for Women in Africa
ATM	Automated Teller Machine
ASCAs	Accumulating Savings and Credit Associations
APY	Afal Pension Yojana
AQIM	AI Qaeda in the Islamic Maghreb
AU	African Union
BNV	Bank Verification Number
BOP	Balance of Payments
CARES	Coronavirus Aid, Relief and Economic Countries
CBK	Central Bank of Kenya
CJTF	Civilian Joint Task Force
COHRE	Centre on Housing Rights and Evictions
DDR	Disarmament, Demobilization and Re-integration
DFI	Digital Financial Inclusion
DHS	Demographic and Health Survey
ECOWAS	Economic Community of West African States
EFINA	Enhancing Financial Innovation and Access
ERGP	Economic Recovery and Growth Plan
FIAP	Financial Inclusion Action Plan
FINTECH	Financial Technologies
FOCCAS	Foundation for Credit and Community Assistance
GBV	Gender-Based Violence
GDP	Gross Domestic Product
GPFI	Global Partnership for Financial Inclusion
GoZ	Government of Zimbabwe
HDI	Human Development Index
IBRD	International Bank for Reconstruction and Development
ICT	Information Communication Technology
ILO	International Labour Organization
ISAL	Income Saving and Lending Groups
ISWA	Islamic State of West Africa
ISWAP	Islamic State West African Province
KTA	Knowledge Transfer Africa
KYC	Know Your Custom
LAPO	Lift Above Poverty
LGA	Local Government Areas

LIC	Low Income Country
MIWE	Mastercard Index of Women Entrepreneurs
MFIs	Micro Finance Institutions
MSMEs	Micro, Small and Medium Enterprises
NPA	National Prosecuting Agency
NDIC	Nigerian Deposit Insurance Corporations
NDS1	National Development Strategy-1
NFIS	National Financial Inclusion Strategy
OAU	Organisation of African Unity
OECD	Organisation for Economic Co-operation and Development
OSC	Operation Safe Corridor
OPEC	Organization for Petroleum Exporting Countries
PAVE	Partnership Against Violent Extremism
PMJDY	Pradhan Mantri Jan Dhan Yojana
PMJJBY	Pradhan Mantri Jeevan Jyoti Bima Yojana
PMSBY	Pradhan Mantri Suraksha Bima Yojana
PSC	Public Service Commission
POS	Point of Sale
RBZ	Reserve Bank of Zimbabwe
RIA	Registered Investment Advisor
RSA	Republic of South Africa
ROSCAs	Rotating Savings and Credit Associations
SANCO	South African National Civics Organisation
SSA	Sub-Saharan Africa
SASSA	South Africa Social Security Agency
SAPS	South African Police Services
SDG	Sustainable Development Goals
SEAP	Self-reliance Economic Advancement Programme
SHGs	Self-Help Groups
SI	Statutory Instrument
SIU	Special Investigating Unit
SMEs	Small and Medium-sized Enterprises
SONA	State of Nation Address
TSP	Transitional Stabilization Plans
UMIC	Upper-Middle Income Country
UN	United Nations
UNCDF	United Nations Capital Development Fund
UNGA	United Nations General Assembly
UNHR	Universal Declaration of Human Rights
UNDP	United Nations Development Programme
UOB	Urwego Opportunity Bank
WAEMU	West African Economic and Monetary Union
WEDTF	Women Entrepreneurship Development Trust Fund
WHO	World Health Organization
WIEGO	Women in Informal Employment: Globalizing and Organizing
ZCIEA	Zimbabwe Chamber of the Informal Economy Association

Contents

List of Figures

List of Tables

Contributors

Temitayo Adeyemo (PhD) is an agricultural economist, with a focus on sustainable and inclusive agrifood systems. Her post-doctoral fellowship in the BiomassWeb project of the Centre for Development Research (ZEF), University of Bonn. She worked in policy research at the Agricultural and Food Policy Department of the Nigerian Institute of Social and Economic Research (NISER). Temitayo is presently a faculty member at the Department of Agricultural Economics, University of Ibadan, where she combines teaching and research. She is a policy fellow in the African Women in Agricultural Research and Development/Gender Responsive Agricultural System Policy (AWARD/GRASP) programme.

Doyinsola Oluwagbemi Agunbiade has a PhD in Peace and Conflict Studies from the Department of Peace, Security and Humanitarian Studies and a master's in Peace and Conflict Studies from the University Ibadan, Nigeria. He obtained a BSc (Hons) in Political Science from the Ambrose Ali University, Ekpoma, Nigeria. He is an African fellow (2023/2024) of the Harry Frank Guggenheim Foundation, New York, USA, as well as a Cohort-1 member of the WARIA (Writing and Researching the Political Inequality of Africa), a British Academy funded programme. He has significantly contributed to research, publications, and scholarship.

Kemi Funlayo Akeju holds a PhD in Economics from Ekiti State University, Ado-Ekiti, Nigeria, with specialization in the field of Development and Health Economics. Her interest is on issues of African development and sustainable agenda with a focus on gender and development studies, poverty, demographic and health, financial inclusion, inequality, climate change, and food security. She has proven experience in developing micro- and macroeconomic models to inform policies to aid strategic plans for advocacy and solution-based research on SDG goals and targets. She has attended and presented academic papers at both local and international conferences.

Yetunde Adebunmi Aluko is a research professor in the Social Policy Department of the Nigerian Institute of Social and Economic Research (NISER). She is a Professor of Sociology with a special interest in gender and development studies. She has worked extensively on research related to promoting Gender Equality and Social Inclusion (GESI) in the workplace and creating opportunities for women to ensure their inclusion in development in general. She is a member of several professional bodies and associations including the International Association for Feminist Economics (IAFFE), among others.

Michelle Beeslaar is the co-owner of Faolan Consulting, a project management controls specialist consultancy with headquarters in Dubai, UAE, with listed and multinational blue-chip clients whose bases or projects on which we work span six continents. Her focus is on business development, research, and innovation undertaking the aspects of work that relate to Environmental and Social Governance (ESG) linked to sustainable corporate investment. She is a PhD candidate (International Relations) at the University of Pretoria. She is a former mid to senior public servant for 16 years. She is alumni of the US Young Council of Political Leaders and Mandela Washington Fellow nominee.

Belinda R. Chaora is the current Coordinator for the Centre for Entrepreneurship and Financial Inclusion at SIVIO Institute. She is a Master of Science graduate of the University of Birmingham (United Kingdom). Her current research interests include livelihoods; financial inclusion; micro, small, and medium enterprises; industrial development frameworks; and sustainable development. She has developed multiple training programmes (off and online) for startup founders in Zimbabwe and Africa with a focus on problem-solving social entrepreneurship aiming to address Africa's most pressing problems through business development. Her work on financial inclusion has seen extensive research into Zimbabwe, Zambia, Malawi, and Kenya focusing on the financial inclusion of micro, small and medium enterprises and the development of indices to measure the state of financial inclusion in African countries. In addition to this, Belinda Chaora is involved in promoting philanthropy through individual and corporate giving for the non-profit sector.

Edson Chiwenga is a doctoral researcher under the Human Economy Research Programme at the University of Pretoria. His research focuses on digital finance, mobile money, and financial inclusion, more specifically on how digital finance is being adopted and experienced in different political ecologies and cultures. He has interests in issues to do with inclusive development and financial inclusion. Edson holds a master's in Sociology and Social Anthropology (University of Zimbabwe), a BSc Hons Sociology (Great Zimbabwe University), and a Diploma in Human Resources Management (IPMZ).

Queensley C. Chukwudum is a senior lecturer at the University of Uyo, Nigeria. She holds a PhD in Financial Mathematics from the Pan African University for Basic Sciences, Technology and Innovation, Kenya (African Union). She is a lecturer at the University of Uyo teaching her research niche on mathematics, statistics, risk management, and computer science.

Purity Hamunakwadi is a post-doctoral research fellow at Nelson Mandela University. She holds a PhD in Social Transformation and Management

Social Sciences from Northwest University. She is a seasoned researcher and a development professional in the international development sector. Her research interest is related to human settlements and housing, development policy, economic empowerment of the marginalized, sustainable livelihoods, food security, poverty alleviation, socio-economic development, sustainability development, community development, corporate social responsibility, small and medium-sized enterprises, mining, immigrants, public policy, civil society organizations, and the NGOs/NPOs sector.

Fadila Jumare holds a PhD in Social Development at the University of Cape Town. She is a research associate at the Busara Center for Behavioral Economics. She is responsible for supporting both the advisory and research arms of Busara by leading the execution of Busara projects. Her research niche is on nutrition, gender, health, and development. She is an expert in designing high-quality qualitative, quantitative, or experimental research, which takes into account key sampling, instrument, research integrity, ethics, analysis, and intervention considerations on a variety of internal and external projects.

Oluwaseyi Kehinde-Peters is a visionary leader and advocate for women's economic empowerment and financial inclusion in Africa. As the Founder and Executive Director of PAWEN, an EdTech charity, she has significantly impacted over 3500 women across 39 African countries. Oluwaseyi has authored position papers on gender equality and financial inclusion, speaking at prestigious platforms. With multiple degrees and entrepreneurship fellowships, her commitment to continuous learning and advocacy for women's empowerment in business and technology distinguishes her as an influential figure in driving financial inclusion across Africa.

Eric Gwandega Magale is a researcher with a background in finance and risk. He researches the complex relationship between finance, livelihoods, and human development. He holds a master's in Development Finance from the University of Cape Town, Graduate School of Business and a PhD in Development Studies from the University of Pretoria and is a Chartered Accountant.

Sijekula Mbanga is an associate professor, a programme leader for Human Settlements Studies, a member of the research group on environmental stewardship and sustainable livelihoods, and the Chair for Sustainable Human Settlements at Nelson Mandela University. He holds a PhD in Public Administration, with his thesis focusing on integrated development planning, obtained from the previous Nelson Mandela Metropolitan University. His research interests include integrated approaches to development management, sustainable livelihoods, spatial transformation and urbanism, green settlements, and neighbourhoods, human settlement policies and systems, local government capacity, and indigenous knowledge systems in infrastructure development.

Frank Moffat holds a PhD in Urban and Regional Planning, one of the accredited professional degrees offered by the University of Venda. His research interests are urban and regional planning, health, spatial transformation, migration, and economics.

Rifkatu Nghargbu is a health economist and a lecturer in the Department of Economics, Nasarawa State University, Keffi. She has a PhD from the University of Ibadan under the African Economic Research Consortium collaborative PhD programme. She is a fellow of the African Population and Health Research Center. Rifkatu Nghargbu has 13 years of working experience as a lecturer and research fellow with Usmanu Danfodiyo University Sokoto; Federal University of Lafia; National Institute for Legislative and Democratic Studies; National Assembly, Abuja; and Nasarawa State University, Keffi.

Blessing Olanrewaju is an economic researcher known for her significant strides in applied and theoretical economic research. Olanrewaju's research interests include long-run economic growth, energy economics, public policy, and development economics.

Omosefe Oyekanmi holds a PhD in Political Science. She is a post-doctoral fellow at the Pan African Women Studies Unit, Institute of Pan African Thought and Conversation, University of Johannesburg. She is also a research fellow at the Nigerian Institute of Social and Economic Research, Ibadan, Nigeria. Omosefe is a recipient of the British Academy funded programme, Writing and Researching the Political Economy of Inequality in Africa (WARIA), cohorts 2021 and 2023, and a prestigious scholar of CODESRIA College of Mentors and Mentees (2019). She has over 30 published articles, across humanitarianism, development, public policy, gender issues, and international peace building.

About the Editor

Tinuade Adekunbi Ojo is an associate professor in the Department of Politics and International Relations, who serves on research purposes; as part of the team leading work on Pan African Women Studies at the Institute of Pan African Thought and Conversation, University of Johannesburg. She is very passionate, energetic, and authentic with her approaches. She is a global transformational keynote speaker, educational consultant, and mentoring coach who impacts others for purpose! She holds a PhD in Political Science from the University of Pretoria. Her master's and honors were in Development Studies from UNISA. As a feminist political economist, her work explores gender and politics, the gendered dimension of trade, financial inclusion, sustainable development goals, digital transformation in policy, politics and digitalization, international political economy, gender inequality, and poverty reduction strategies. Tinuade has published a wide range of works on gender and development policy and the IMF's external relations with Africa, BRICS, Southern Africa, and West Africa, focusing on the political economy of trade and aid. In addition, she has acted as a peer reviewer for renowned scientific journals.

Additionally, Tinuade has held and still holds various administrative portfolios, including academic, supervisor, usher, mentor, and administrator for postgraduate students. She was a former recipient and is currently a co-facilitator of the British Academy-funded programme Writing and Researching the Political Economy of Inequality in Africa (WARIA). She is an associate editor for the *African Journal for Political*

Science (AJPS) and journal and managing editor for *BRICS*, and researcher for the Digital Policy Studies Unit (4DPRU) at Johannesburg. Tinuade is a member of international and local associates, CAAS, IPSA, AAPS, ASA, DSA, UNASA UP, SACE, SAAPS, and others. Tinuade was the African Association of Political Science (AAPS) program officer. She was also responsible for Partnership and Resource Mobilization for CBW-Africa and has received many awards, scholarships and grants, presented at local and international conferences.Tinuade has also worked with several civil societies and international institutions to advance the course on women's economic empowerment and sustainability. Her recent work debates: Ojo, TA. (2023). "The Politics of Financial Inclusion in South Africa: Evolution and Lessons". Springer. https://link.springer.com/book/9789819918461#book-header Released July 23rd 2023. Hardcopy ISBN 978-981-99-1846-1, e-Book ISBN 978-981-99-1847-8; and Ojo, T.A. (2023). Exploring Digital and Financial Divide and Its Effects on South African Women's Engagement in Entrepreneurship. In: Osman, A., Nagle, J., Tripathi, S. (eds) The Urban Ecologies of Divided Cities. UEDC 2022. Advances in Science, Technology & Innovation. Springer, Cham. https://doi.org/10.1007/978-3-031-27308-7_24.

Gendered Finance: Inclusion and Transformation

1

Tinuade Adekunbi Ojo

Abstract

Gendered finance in Africa represents a multifaceted landscape where financial systems intersect with gender dynamics, presenting both challenges and opportunities for inclusive economic growth. This abstract explores the complexities surrounding gendered finance in Africa, emphasizing the pivotal role of inclusion and transformation in reshaping financial structures. The African continent grapples with entrenched gender disparities in access to financial services, with women often facing significant hurdles in obtaining credit, owning property, or participating in formal financial systems. These barriers are compounded by cultural norms, limited access to education, and discriminatory practices, hindering women's economic empowerment and financial independence.

However, amidst these challenges lie prospects for transformative change. Efforts aimed at fostering inclusive financial ecosystems are gaining momentum. Initiatives promoting financial literacy, access to microfinance, and innovative fintech solutions tailored to women's needs are emerging across the continent. Moreover, policy interventions and advocacy for gender-sensitive regulations are gradually reshaping the financial landscape. Understanding the intersectionality of gender, finance, and socioeconomic contexts is crucial in devising sustainable strategies for inclusive growth. Empowering women economically not only enhances their individual agency but also contributes significantly to broader societal development, fostering economic resilience and reducing poverty.

This abstract underscores the imperative of addressing gendered financial inequalities in Africa, emphasizing the need for holistic approaches that combine policy reforms, innovative financial products, and cultural shifts. By embracing inclusion and transformation in financial systems, Africa can unlock the immense potential embedded in its diverse population, fostering a more equitable and prosperous future for all.

Keywords

Financial inclusion · Gender dynamics · African economies · Transformation · Empowerment

T. A. Ojo (✉)
Department of Politics and International Relations,
University of Johannesburg,
Johannesburg, South Africa

T. A. Ojo (ed.), *Women and Finance in Africa*, Sustainable Development Goals Series,
https://doi.org/10.1007/978-3-031-53337-2_1

1.1 Introduction

In the realm of economic development, the concept of gendered finance has emerged as a pivotal area of focus, particularly within the context of fostering financial inclusion in Africa (Gabor and Brooks 2020). At its core, gendered finance encapsulates the nuanced intersection of gender dynamics and financial systems, specifically emphasizing the unique challenges, opportunities, and implications for women's access to and utilization of financial services (Haynes 2017). Within the African landscape, this facet of finance bears immense significance, shaping not only individual economic empowerment but also influencing broader societal and developmental trajectories (Bouman 2019). The pressing importance of addressing gender disparities in financial access cannot be overstated. Across diverse African regions, there exists a profound asymmetry in access to financial resources between genders, with women disproportionately affected by limited access to formal banking services, credit facilities, and other fundamental financial instruments. This imbalance is multifaceted, stemming from a confluence of cultural, economic, and institutional barriers that collectively impede women's financial inclusion. Over the past 30 years, considerable progress has been made, marked by significant agreements like the Beijing Declaration, the Convention on the Elimination of all Forms of Discrimination Against Women (CEDAW), the 2030 Agenda for Sustainable Development, including the dedicated Sustainable Development Goal (SDG) 5 and the integration of gender-sensitive indicators into various other SDGs and the African Union's Agenda 2063 (UNECA 2023). The implementation of these agreements has resulted in measurable advancements in Africa, including noteworthy reductions in maternal mortality, heightened advocacy against violence toward women, and legislative efforts securing women's rights to education and inheritance.

However, despite these advancements, gender disparities persist. Women still face human rights violations, limited access to education, healthcare, employment, decision-making roles, leadership positions, and economic resources (UNECA 2023). Challenges are compounded by factors such as infectious disease outbreaks (like the ongoing COVID-19 pandemic) and climate change, further exacerbating existing inequalities. Addressing violence against women remains a critical and urgent issue.

This chapter endeavors to navigate the intricate tapestry of gendered finance in Africa by delineating the multifaceted dimensions of this issue and exploring its ramifications. Through a comprehensive exploration, it seeks to underscore the imperatives of addressing gender disparities in financial access, recognizing the pivotal role of inclusive financial systems in driving sustainable development and socioeconomic progress.

The structure of this chapter is designed to traverse through the various dimensions of gendered finance. It begins by elucidating the concept and significance of gendered finance within the realm of financial inclusion in Africa. Subsequently, it delves into an exploration of the prevailing disparities in financial access, unraveling the multifarious factors contributing to these inequities. Additionally, the chapter highlights the transformative potential of enhancing women's financial inclusion, emphasizing its multifaceted impact on individual empowerment, economic development, and societal well-being. By providing a comprehensive overview of existing initiatives, the chapter evaluates ongoing programs aimed at mitigating gender disparities in financial access. Furthermore, it critically examines the role of technological innovations and policy interventions in ameliorating these disparities, emphasizing their potential to catalyze transformative change.

In essence, this chapter seeks to weave together a fabric of knowledge and insights, shedding light on the intricate nexus between gender, finance, and inclusive development in Africa. Through an academic exploration of these themes, it aims to contribute to a deeper understanding of the challenges and opportunities inherent in advancing gendered finance for inclusive growth in the African context.

1.2 Gender Disparities in Financial Access in Africa

Financial inclusion remains a critical challenge across numerous African nations, significantly affecting women, who encounter pronounced barriers in accessing and utilizing formal financial services. Statistical analysis reveals a persistent gender gap in financial inclusion across the continent. According to recent studies released by World Bank and the African Development Bank, there is a gender gap of approximately 21% from 2021 Findex Report of women in Africa lack access to formal financial services compared to their male counterparts (Ferrer et al. 2023). This gender disparity is evident across various dimensions of financial inclusion, encompassing savings, credit, insurance, and other financial products.

Several interwoven factors contribute to these disparities. Cultural norms and societal perceptions often dictate women's roles in economic activities, relegating them to household duties and informal sectors, limiting their financial independence (Khan et al. 2023). Economic disparities exacerbate these challenges, with women disproportionately affected by poverty, lack of assets, and limited employment opportunities, constraining their ability to access formal financial services (Khan et al. 2023). Legal barriers, including discriminatory inheritance laws and property rights, further impede women's financial inclusion, undermining their ability to leverage financial resources. Case studies across diverse African regions illuminate specific challenges faced by women in accessing financial services. In rural areas of Nigeria, for instance, cultural norms and limited financial literacy perpetuate a cycle of financial exclusion among women (Onoshakpor et al. 2023). These norms often restrict women's mobility, hindering their access to banks or financial institutions. Similarly, in countries like Kenya and Tanzania, the lack of collateral and credit history disproportionately affects women entrepreneurs seeking loans for business ventures (Matafwali 2022). Moreover, in conflict-affected regions such as South Sudan or the Democratic Republic of Congo, women encounter heightened challenges accessing financial services due to disrupted infrastructures and increased vulnerability to economic shocks.

Furthermore, microfinance initiatives, often heralded as avenues for financial inclusion, sometimes fall short of addressing the multifaceted barriers faced by women. While these initiatives aim to provide access to credit and savings, they might overlook the broader structural impediments, including limited access to education and healthcare, which profoundly impact women's economic agency. The complex web of challenges confronting women in accessing financial services necessitates a multifaceted approach that addresses cultural norms, economic disparities, legal frameworks, and institutional barriers. Strategies aimed at promoting gender-sensitive financial products, enhancing financial literacy among women, and advocating for policy reforms to address legal disparities are imperative to bridge the gender gap in financial inclusion.

1.3 Importance of Financial Inclusion for Women

Financial inclusion stands as a cornerstone for fostering empowerment and economic autonomy among women, transcending individual benefits to encompass broader societal advancements (Ojo 2021). Enabling women's access to and utilization of financial services not only catalyzes economic growth but also contributes significantly to gender equality and sustainable development agendas (Prabhakar 2019). The following section highlights the importance of financial inclusion.

1.3.1 Empowerment and Economic Benefits of Financial Inclusion

Financial inclusion serves as a catalyst for women's empowerment, providing them with the tools and resources to exercise greater control over their economic lives (Prabhakar 2019). Access to formal financial services, including savings accounts, credit facilities, and insurance, empowers women to make independent financial decisions, fostering self-reliance and reducing their vulnerability to economic shocks (Prabhakar 2019). Moreover, financial inclusion facilitates women's participation in economic activities, enabling them to contribute actively to household incomes and the overall economy.

1.3.2 Impact on Household Income, Education, Health, and Entrepreneurship

The ramifications of financial inclusion also extend beyond individual empowerment, manifesting in enhanced household welfare, improved education, better health outcomes, and increased entrepreneurship among women (Pal and Gupta 2023). Studies indicate a positive correlation between women's access to financial services and household income. When women have control over financial resources, they tend to allocate a higher proportion toward children's education and healthcare, thereby improving overall family well-being (Pal and Gupta 2023). Furthermore, financial inclusion serves as a catalyst for women's entrepreneurship. Access to credit and savings allows women to start or expand businesses, thereby contributing to job creation and economic growth (Andriamahery and Qamruzzaman 2022). This entrepreneurial engagement not only boosts individual incomes but also engenders innovation and diversification within economies.

1.3.3 Linkages Between Gender Equality and Economic Development

Gender equality and economic development are intrinsically interconnected, with financial inclusion serving as a linchpin for achieving both (Ojo 2023). Empowering women economically through enhanced financial access and control over resources fuels their participation in the formal economy, contributing to increased productivity and economic diversification. Closing the gender gap in financial inclusion can potentially unlock substantial economic gains, thereby propelling overall economic development (Ojo 2023).

Moreover, encouraging gender equality in financial access and decision-making processes fosters social inclusion and stability (Fernández et al. 2021). Women with financial independence are better positioned to engage in civic and community activities, amplifying their voices in social and political spheres, consequently influencing policy directions toward more inclusive and equitable outcomes (Pal and Gupta 2023).

1.4 Existing Initiatives and Programs for Gendered Financial Inclusion in Africa

The pursuit of gendered financial inclusion in Africa has seen the emergence of diverse programs and initiatives orchestrated by governments, international organizations, financial institutions, and civil society entities (Kayongo et al. 2021). These endeavors aim to bridge the gender gap in financial access, foster women's economic empowerment, and contribute to broader socioeconomic development (Mader et al. 2022).

Overview of Programs

Initiatives such as the African Development Bank's Affirmative Finance Action for Women in Africa (AFAWA) and the United Nations' Economic Commission for Africa's Women's Financial and Economic Inclusion Initiative (WFEII) stand as prominent endeavors (Buvinic et al. 2020; United Nations Economic Commission for Africa (UNECA) 2023). These programs strive to enhance women's access to finance by providing targeted financial products, capacity-building initiatives, and policy advocacy efforts.

The United Nations Economic Commission for Africa (UNECA) is committed to advancing gender equality and empowering women across its 54 African member states (UNECA 2023). At the core of this commitment lies the gender equality and women's empowerment section within the organization, dedicated to addressing persistent gender disparities in crucial sectors such as digital finance, agriculture, and trade. Its focus is to drive economic growth by leveraging diverse expertise within UNECA and executing transformative actions to address gender-related issues (UNECA 2023). These initiatives involve extensive research, technical support, and policy dialogues aimed at breaking barriers to gender equality.

To achieve this aim, UNECA provides tailored operational support to each member state, employing sustainable approaches to aid in fulfilling their gender equality commitments. This support involves enhancing policy and program designs, boosting implementation capacity, and strengthening institutional development (UNECA 2023). It includes tailored policy tools, exchange of best practices at various levels, research and feasibility studies, targeted capacity-building efforts, and national reviews to monitor the implementation of international agreements.

Convening activities play a pivotal role in bringing together representatives from governments, academia, civil society, and private sector to discuss challenges, explore policies, foster partnerships, and build regional consensus. These activities are essential platforms for coordinating action on regional and international agreements that require cross-border collaboration. Key events encompass the Africa Regional Consultations on the Commission on the Status of Women, the statutory meeting of the Committee on Social Policy, Poverty, and Gender, and the Africa regional review of the Beijing Declaration and Platform for Action.

UNECA's commitment to supporting member states in achieving gender equality and women's empowerment is resolute, aligning with the organization's mandates and programs. Attaining these objectives is expected to yield substantial benefits for families, societies, and economies, serving as the linchpin for realizing all the goals of the 2030 Agenda for Sustainable Development.

Furthermore, the African Development Bank's Affirmative Finance Action for Women in Africa (AFAWA) initiative plays a pivotal role in addressing the gender gap in financial access and unleashing the entrepreneurial potential of women across the continent (African Development Bank Group (AfDB) 2024). AFAWA operates through three primary pillars:

1. *AFAWA finance*: Leveraging the African Development Bank's financial instruments, this pillar employs innovative solutions to revolutionize the banking and financial landscape. It includes a $300 million risk-sharing instrument to facilitate $3 billion in credit for women-owned businesses in Africa. Additionally, a rating system evaluates financial institutions based on their lending to women and socioeconomic impact, rewarding top-performing institutions with preferential financing terms from the African Development Bank (AfDB 2024).
2. *AFAWA technical assistance*: This pillar provides advisory services to financial institutions, ensuring the successful implementation of their product portfolios for women. It also strengthens the capacity of women entrepreneurs through training aimed at enhancing business productivity and growth (AfDB 2024).
3. *AFAWA enabling environment*: Engaging with African governments and key stakeholders, this pillar supports legal, policy, and regulatory reforms aimed at eliminating structural barriers hindering women in business (AfDB 2024).

These combined efforts of UNECA and AFAWA represent critical steps toward realizing gender equality and empowering women across Africa, fostering sustainable development and economic growth. Moreover, various microfinance institutions and NGOs across African nations have pioneered initiatives tailored to women's financial inclusion (Ukala et al. 2020). These initiatives often encompass financial literacy programs, microcredit schemes, and the provision of savings and insurance products designed specifically for women.

Evaluation of Effectiveness and Implementation Challenges

While these initiatives signify significant strides toward gendered financial inclusion, their effectiveness faces multifaceted challenges (UNECA 2023). One major hurdle is the persistent gender norms and sociocultural barriers that constrain women's participation in formal financial systems. Deep-seated societal perceptions often limit women's financial agency, hindering their uptake of financial services despite the availability of programs. Inadequate access to resources, including limited collateral and credit history, poses a significant barrier, especially for women entrepreneurs seeking loans to start or expand businesses (UNECA 2023). Furthermore, challenges in scaling up these programs to reach remote and marginalized communities exacerbate implementation hurdles, often due to infrastructural constraints and geographical disparities.

Examples of Successful Projects and Empowerment Impact

Notwithstanding these challenges, several initiatives have demonstrated remarkable success in empowering women through financial inclusion. Case studies from diverse African nations highlight instances where these programs have facilitated women's economic independence and empowerment (UNECA 2023). For instance, microfinance initiatives like the Grameen Foundation's Village Savings and Loan Associations (VSLAs) in Uganda have empowered rural women by providing access to small loans and fostering financial literacy, leading to increased savings and entrepreneurial activities (Chilongozi 2022).

Similarly, AFAWA's initiatives in countries like Rwanda and Senegal have facilitated women's access to credit and training, resulting in a surge in women-led businesses and improved household incomes (Buvinic et al. 2020). The impact of these programs extends beyond economic realms, empowering women to assert their rights and roles within their households and communities.

1.5 Role of Technology and Innovation in Bridging the Gender Gap in Financial Access

Technology, especially financial technology (fintech), has emerged as a powerful enabler in addressing the gender gap in financial access across the African continent (Ojo 2022). Fintech solutions offer innovative approaches to enhance women's participation in formal financial systems, providing tailored products and services to overcome traditional barriers (Ojo 2022).

Fintech's Role in Bridging the Gender Gap

Fintech innovations play a pivotal role in mitigating the gender gap by providing convenient, accessible, and user-friendly financial services. Mobile money platforms, such as M-Pesa in Kenya and EcoCash in Zimbabwe, have revolutionized financial inclusion by offering digital payment solutions, enabling women, even in remote areas, to engage in financial transactions without the need for a traditional bank account (Muzanechita 2022; Mpofu 2022). Furthermore, fintech applications offering financial education and budgeting tools specifically designed for women have contributed to enhancing financial literacy and empowering women to make informed financial decisions (Mpofu 2022). These platforms often incorporate localized content and interactive interfaces, catering to diverse cultural contexts and literacy levels.

Case Studies on Innovative Solutions

In African countries like Nigeria and Ghana, fintech startups have introduced innovative solutions targeting women's financial inclusion (Danladi et al. 2023). For instance, apps like "Shecluded" in Nigeria provide financial management tools, investment opportunities, and networking spaces exclusively for women, fostering a supportive ecosystem for financial growth and empowerment (Financial Alliance for Women 2022). Similarly, in Kenya, initiatives such as "M-Shwari" have revolutionized lending by offering micro-loans directly through mobile phones, eliminating traditional barriers such as collateral requirements and lengthy application processes (Kiptorus 2019). These initiatives have significantly expanded access to credit for women entrepreneurs, enabling them to invest in and grow their businesses.

Challenges and Opportunities in Technological Advancements

Despite the promising potential of fintech, challenges persist in ensuring equitable access and maximizing the benefits for women (Danladi et al. 2023). Gender-based digital divides, including disparities in mobile phone ownership, Internet access, and digital literacy, remain significant barriers, particularly in rural and marginalized communities (Sterling et al. 2020). Moreover, concerns regarding data privacy, security, and trust in digital financial services pose challenges that need to be addressed to ensure women's confidence in using these platforms (Rana et al. 2020). However, technological advancements present vast opportunities. Blockchain technology, for instance, offers transparent and secure financial transactions, potentially reducing administrative barriers and enhancing trust in financial systems (Chen and Bellavitis 2019). Moreover, partnerships between fintech firms, financial institutions, and regulatory bodies can create synergies to develop more inclusive and gender-responsive financial solutions.

1.6 Policy Recommendations and Way Forward for Gendered Financial Inclusion

Policy Suggestions for Governments and Regulatory Bodies

Governments and regulatory bodies play a pivotal role in fostering an enabling environment for gendered financial inclusion (Natile 2020). To enhance this, policymakers should prioritize the formulation and implementation of policies that promote gender equality in financial access. This includes:

(a) *Legal reforms*: Implementing legal reforms to address discriminatory laws that limit women's access to property rights, inheritance, and financial services (Christopherson et al. 2022). Ensuring equal rights to assets and property ownership empowers women economically.

(b) *Financial inclusion mandates*: Enforcing financial inclusion mandates that specifically target women, requiring financial institutions to allocate a certain percentage of their services and products toward women-centric initiatives Sathiyabama et al. 2023).

(c) *Supportive infrastructure*: Investing in infrastructure, such as reliable Internet connectivity and mobile network coverage, to bridge the digital divide and ensure equitable access to financial services, especially in rural and underserved areas (Reddick et al. 2020).

Strategies for Financial Institutions

Financial institutions can contribute significantly to gendered financial inclusion by designing and offering gender-sensitive products and services that cater to the unique needs of women (Ojo and Zondi 2021). Strategies include:

(i) *Tailored financial products*: Developing financial products specifically designed for women, considering their financial goals, risk tolerance, and preferences (Bapat 2020). This includes savings accounts, credit facilities with flexible repayment terms, and insurance products covering specific women-centric needs.

(ii) *Financial literacy programs*: Providing comprehensive financial literacy programs targeted at women to enhance their knowledge and confidence in managing financial matters. These programs should cover basic financial concepts, investment opportunities, and entrepreneurship skills (Ojo 2020).

(iii) *Diverse representation*: Promoting gender diversity within financial institutions'leadership and decision-making processes to ensure a more inclusive approach in designing and implementing financial products and services (Gallego-Sosa et al. 2021).

Importance of Partnerships Between Stakeholders
Collaboration among diverse stakeholders is critical for sustained progress in gendered financial inclusion (Ojo 2021). Partnerships between governments, financial institutions, civil society organizations, and international agencies are imperative for:

(a) *Knowledge sharing and capacity building*: Facilitating knowledge sharing and capacity-building initiatives among stakeholders to exchange best practices, innovative ideas, and lessons learned in advancing gendered financial inclusion (Jackson and de Morais Sarmento 2020).
(b) *Resource mobilization*: Pooling resources and expertise to scale up successful initiatives and pilot projects, reaching a larger segment of the population, especially in underserved regions (Rey-Moreno et al. 2021).
(c) *Advocacy and awareness*: Engaging in joint advocacy efforts to raise awareness about the importance of gendered financial inclusion and mobilize support for policy reforms and initiatives (Natile 2020).

1.7 Conclusion

The significance of gendered finance in shaping African development cannot be overstated. Gendered finance stands as a linchpin for inclusive growth, economic empowerment, and societal progress, underlining the imperative need to address gender disparities in financial access across the continent. Gendered finance also serves as a catalyst for inclusive and sustainable development in Africa. It not only unlocks the economic potential of women but also fosters broader socioeconomic advancements. Empowering women through enhanced financial access and tailored financial services contributes significantly to poverty alleviation, economic growth, and societal well-being. Moreover, it strengthens the fabric of communities by promoting gender equality and empowering women as agents of change.

Throughout this exploration, several critical findings have emerged. The persistent gender gap in financial inclusion across various African regions highlights the urgent need for concerted efforts. Cultural, economic, and legal barriers continue to impede women's access to formal financial services, necessitating multifaceted strategies to address these challenges. Recommendations include legal reforms to ensure equal rights, policies mandating financial inclusion for women, tailored financial products by institutions, and collaborative partnerships among stakeholders. Emphasizing financial literacy programs, leveraging technology, and fostering diverse representation are pivotal for sustained progress. The time for action is now. Stakeholders across governments, financial institutions, civil society organizations, and international bodies must prioritize gendered financial inclusion as a fundamental component of development agendas. It necessitates a collective commitment to implementing policy reforms, designing inclusive financial products, fostering partnerships, and investing in capacity building. By prioritizing gendered finance, stakeholders can unlock immense potential, not only transforming the lives of individual women but also catalyzing broader socioeconomic advancements, fostering resilience, and driving inclusive growth across the African continent.

References

African Development Bank Group (AfDB) (2024) Affirmative Finance Action for Women in Africa (AFAWA). https://www.afdb.org/en/topics-and-sectors/initiatives-partnerships/afawa-affirmative-finance-action-for-women-in-africa/what-does-afawa-do

Andriamahery A, Qamruzzaman M (2022) Do access to finance, technical know-how, and financial literacy offer women empowerment through women's entrepreneurial development? Front Psychol 12:776844

Bapat D (2020) Antecedents to responsible financial management behavior among young adults: moderating role of financial risk tolerance. Int J Bank Mark 38(5):1177–1194

Bouman FJ (ed) (2019) Financial landscapes reconstructed: the fine art of mapping development. Routledge

Buvinic, M., O'Donnell, M., & Bourgault, S. (2020). Women's economic empowerment in West Africa: towards a practical research agenda. Center for Global Development Working Paper, 557

Chen, Y., & Bellavitis, C. (2019). Decentralized finance: blockchain technology and the quest for an open financial system. Stevens Institute of Technology School of Business Research Paper

Chilongozi MN (2022) Microfinance as a tool for socio-economic empowerment of rural women in northern Malawi: a practical theological reflection. Doctoral thesis. Stellenbosh University

Christopherson K, Yiadom A, Johnson J, Fernando F, Yazid H, Thiemann C (2022) Tackling legal impediments to women's economic empowerment. International Monetary Fund, 2022, 1

Danladi S, Prasad MSV, Modibbo UM, Ahmadi SA, Ghasemi P (2023) Attaining sustainable development goals through financial inclusion: exploring collaborative approaches to Fintech adoption in developing economies. Sustain For 15(17):13039

Fernández R, Isakova A, Luna F, Rambousek B (2021) Gender equality and inclusive growth. International Monetary Fund

Ferrer M, Perrin C, Jacolin L (2023) Sub-Sahara Africa: the fimancial gender gap between women and men. https://blogs.worldbank.org/allaboutfinance/sub-saharan-africa-financial-gender-gap-between-men-and-women

Financial Alliance for Women (2022) "Shecluded". Available https://financialallianceforwomen.org/members/shecluded/. Accessed 02/01/2023

Gabor D, Brooks S (2020) The digital revolution in financial inclusion: international development in the fintech era. In: Material cultures of Financialisation. Routledge, pp 69–82

Gallego-Sosa C, Gutiérrez-Fernández M, Fernández-Torres Y, Nevado-Gil MT (2021) Corporate social responsibility in the European banking sector: commitment to the 2030 agenda and its relationship with gender diversity. Sustain For 13(4):1731

Haynes K (2017) Accounting as gendering and gendered: a review of 25 years of critical accounting research on gender. Crit Perspect Account 43:110–124

Jackson ET, de Morais Sarmento E (2020) Gender lens investing: co-creating critical knowledge to build a credible, durable field. Global handbook of impact investing: solving global problems via smarter capital markets towards a more sustainable society

Kayongo S, Tom M, Mathiassen L (2021) Organizing and orchestrating microfinance initiatives: a contextual inquiry. Int J Soc Econ 48(2):221–239

Khan NR, Arif GM, Sabir M, Abbasi JY (2023) Unpaid work and women empowerment. J Hum Behav Soc Environ:1–20

Kiptorus JJ (2019) Digital financial inclusion: determinants of M-Shwari in Kenya. Faculty of Commerce, Graduate School of Business (GSB). http://hdl.handle.net/11427/32330

Mader P, Duvendack M, Macdonald K (2022) Fintech and tax in sub-Saharan Africa: taxation versus financial inclusion. J Cult Econ 15(4):488–507

Matafwali K (2022) Investigating into the effects of business credit availability on Sme's.; a case study of Lusaka. Doctoral dissertation, Cavendish University

Mpofu FY (2022) Industry 4.0 in financial services: mobile money taxes, revenue mobilisation, financial inclusion, and the realisation of sustainable development goals (SDGs) in Africa. Sustain For 14(14):8667

Muzanechita RS (2022) The significance of the Ecocash mobile money service in fostering financial inclusion in the Zimbabwean informal sector. Masters dissertation. University of Pretoria

Natile S (2020) The exclusionary politics of digital financial inclusion: mobile money, gendered walls. Routledge, London

Ojo TA (2020) A study of financial inclusion and women's empowerment in South Africa: the case of female entrepreneurs in gauteng (Doctoral dissertation, University of Pretoria (South Africa))

Ojo TA (2021) Gendered finance and economic opportunities in SADC communities: an overview of challenges and prospects. International Journal of African Renaissance Studies 16(2):97–118

Ojo TA (2022) Digital financial inclusion for women in the fourth industrial revolution: a key towards achieving sustainable development goal 5. Africa Review 14(1):98–123

Ojo TA (2023) The politics of financial inclusion of women in South Africa: evolution and lessons. Springer Nature

Ojo TA, Zondi S (2021) Impact of institutional quality and governance on financial inclusion for women in South Africa: a case of Gauteng women entrepreneurs. The Strategic Review for Southern Africa 43(2):59–82

Onoshakpor C, Cunningham J, Gammie E (2023) Contextualising female entrepreneurship and financial inclusion in Nigeria. In: Contextualising African studies: challenges and the way forward. Emerald Publishing Limited, pp 13–36

Pal M, Gupta H (2023) Sustainable women empowerment at the bottom of the pyramid through credit access. Equality, Diversity and Inclusion: An International Journal 42(1):157–171

Prabhakar R (2019) Financial inclusion: a tale of two literatures article accepted for publication in social policy and society. Open Res Online 18(1):37–50

Rana NP, Luthra S, Rao HR (2020) Key challenges to digital financial services in emerging economies: the Indian context. Inf Technol People 33(1):198–229

Reddick CG, Enriquez R, Harris RJ, Sharma B (2020) Determinants of broadband access and affordability: an analysis of a community survey on the digital divide. Cities 106:102904

Rey-Moreno C, Bloom P, Diga K, Jensen M, Ramos KV, Brock N, Baca26, C, Belur27, S.B. (2021) Funding bottom up connectivity: approaches and challenges of community networks to sustain themselves (Chapter 3). In: Belli L, Hadzic S (eds) Community networks: towards sustainable funding models. University Denmark, Aalborg, pp 53–72

Sathiyabama MV, Vidwakalyani R, Priyadharshini BI, Kiruthika T, Ragaprabha MM (2023) Women empowerment through financial inclusion. Shanlax Publications, India

Sterling R, Grubbs L, Koutsky T (2020, December) Breaking through the gender digital divide: technology, social norms, and the WomenConnect challenge. In: TPRC48: the 48th research conference on communication. Information and Internet Policy

Ukala E, Abril ER, Dassanou E (2020) Transformative policy solutions to support women-led businesses in Africa in a post covid-19 world. Policy brief by Efe Ukala, Founder of ImpactHER; Elena Ruiz Abril, Women's Economic Empowerment Regional Advisor, UN Women; Esther Dassanou, Manager, Affirmative Finance Action for Women in Africa, African Development Bank. Accessed 02/01/2023. https://www.developmentaid.org/api/frontend/cms/file/2021/01/policy_brief_-_transformative_policy_solutions.pdf

United Nations Economic Commission for Africa [UNECA] (2023) Gender equality and the economic empowerment of women. United Nations Economic Commission for Africa. Available https://www.uneca.org/gender-equality-and-empowerment-women

2 Informal Financial Services in the Financial Inclusion Matrix: Gendered Perspectives from Kenya and Zimbabwe

Eric Gwandega Magale and Edson Chiwenga

Abstract

The prominence of informal financial services in Kenya, Zimbabwe, and indeed most of Africa is undeniable. The use of informal financial services such as table banking and local savings groups has long been a feature in the financial lives of many Africans, and is particularly popular among women who have historically had to devise ways around cultural and patriarchal structures erected in many spheres, including finance. Informal financial services are also prominent in rural settings where financial inclusion has long been a challenge owing to geographical and infrastructural challenges and the supposed high cost of serving small savers. In the global scene, powerful proponents of financial inclusion frame it in terms of formal financial services. For this reason and others, financial exclusion is deemed a particular problem in Africa despite the rampant use of informal financial services; higher-up proponents of financial inclusion would rather have it that financially excluded women abandon informal services which they are much familiar with and ascend to bank accounts and other formal financial services which they often mistrust or may be inconvenient for their purposes. While acknowledging some of the arguments put forward in the literature about the quality and adequacy of informal financial services, their enduring dominance remains irrefutable. This chapter therefore takes a step back to assess the confluence of informal financial services in Africa and the global discourse on financial inclusion with particular reference to women. The chapter argues that informal financial services continue to play a salient role among women even in the face of rapid adoption of financial technology and a fast-developing financial industry in the continent. The chapter also argues for a broader understanding of financial inclusion with reference to women in Africa and the exploration of synergetic opportunities between informal and formal financial structures as a path to achieving true financial inclusion for women.

Keywords

Informal financial services · Financial inclusion · Gendered perspectives · Kenya · Zimbabwe

E. G. Magale (✉)
Centre for the Advancement of Scholarship, University of Pretoria, Pretoria, South Africa

E. Chiwenga
Human Economy Programme, University of Pretoria, Pretoria, South Africa

T. A. Ojo (ed.), *Women and Finance in Africa*, Sustainable Development Goals Series,
https://doi.org/10.1007/978-3-031-53337-2_2

2.1 Introduction

The discourse on financial inclusion has over the last two decades grown in its influence and has received wide attention in both scholarship and praxis. Financial inclusion is framed as the provision of an array of financial services by formal and to a lesser extent semi-formal institutions to people who do not have access to them. Formal institutions in this case relate to institutions that are either prudentially or non-prudentially regulated and typically include commercial banks, microfinance institutions (MFIs), and financial cooperatives. Semi-formal institutions refer to institutions that are formally registered but are unregulated; institutions belonging to this category vary across jurisdictions but may include cooperative societies, credit-only institutions, and digital lenders, among others. It must be emphasized at the outset that formal, semi-formal, and informal are not natural categories but rather mostly based on legal reasoning.

Within the "formalized" framing of financial inclusion, one is considered to be financially included if they own an account with a formal financial inclusion or a mobile money operator. More recently, the metrics of financial inclusion have been broadened to include the usage of these accounts as a proxy indicator of affordability and quality of the financial services in question. Conspicuously, financial services which are informally provided have been subordinated, indeed summarily excluded, from the financial inclusion discourse (Johnson and Nino-Zarazua 2011). This chapter explores the linkage between informal financial services and the broader subject of financial inclusion.

Financial services sectors in Africa are invariably dualistic in that they consist of formal and informal financial services providers who coexist side by side (Atieno 2001; Zins and Weill 2016). This simultaneity is due to the fact that different service providers provide differentiated financial services to clients who bear distinct characteristics (Atieno 2001). Notably, formal financial service providers typically concentrate on serving the wealthier class, whom they consider more profitable (Prahalad 2002; Beck and Demirgüç-Kunt 2008). For example, Kendall and Voorhies (2014) note that banks ordinarily invest in infrastructure such as automatic teller machines (ATMs), physical offices, and personnel almost exclusively in wealthier, denser (and therefore more profitable) areas. This view, however, fails to consider the growing importance of the informal economy among the ultra-poor, which Prahalad and Hammond (2002) estimate accounts for 40–60% of all economic activity in developing countries.

Informal financial services are fairly wide-ranging and include credit from shopkeepers and other (informal) supply chain creditors, local moneylenders, and social networks which include friends and relatives. By far, the most prominent informal financial service providers in the developing world, which this chapter leverages, are local saving groups common as table baking groups, accumulating savings and credit associations (ASCAs), or rotating savings and credit associations (ROSCAs). These informal financial services pervade the daily lives of people in the developing world, more particularly women. An African Development Bank study by Morsy et al. (2019) points out that women in Africa are much more likely to save and invest informally than their male counterparts. The salience of informal financial services to the women's financial inclusion project, particularly in the developing world where financial exclusion is thought to be exacerbated, therefore, deserves some merit.

In this chapter, we draw attention to the significance of informal financial services as it relates to the extension of financial services to excluded and disenfranchised women. We argue that the decoupling of financial services, within the context of the financial inclusion discourse, into formal and informal services prejudices formal financial services in the fashioning of financial inclusion initiatives and strategies. This narrow conception of financial inclusion, we argue, precludes the delivery of financial inclusion benefits to the people at the margins which would otherwise be realized if informal financial services were properly capacitated and seen as part and parcel of the financial inclusion project. We develop this argument into four main subsequent sections; the first section explores arguments for and against informal financial services. The following two sections examine studies from Zimbabwe and Kenya while the last section strings together key inferences from the case studies which anchors our core argument.

2.2 Debates on Informal Financial Services

Before zooming in on two critical case studies, engaging with competing views advanced in the literature helps in understanding the benefits and drawbacks of informal financial services, starting with the former. Informal financial services are fashioned to bypass the formal financial services which are regarded are inflexible and not properly suited to the needs of average consumers and are expensive. For example, to obtain formal credit, average consumers go through various hoops including the need to supply identity and other documentation. However, the World Bank (2019) notes that in low-income countries (LICs), nearly 40% of adults, most of whom are women lack identity documents (IDs), with sub-Saharan Africa having the largest ID coverage gap. This precludes many women from accessing formal financial services. Such systemic limitations on the access to formal financial services necessitate the devising of informal means of meeting people's financial needs.

Concerning accessing formal credit, a borrower often has to show proof of formal employment as a demonstration of one's capacity to borrow, credit reports to demonstrate their ability to repay the loan, and in most cases, the requirement for security for loans in the form of physical assets. However, many women do not possess such assets, and in many cases, such family assets are likely to be in the name of a male head of the household. For the few women who qualify to access formal credit, they are often confronted by restrictive loan covenants and high borrowing costs. To overcome these constrictions, women turn to shopkeepers, social networks, and local savings groups for small, short-term credit. This is true for many people at the economic margins who almost entirely constitute the financially excluded. Zins and Weill (2016) point out that informal credit is crucial on the African continent given that 41% of individuals borrow from an informal source with the first source being "family and friends" (37.5%) followed by credit from a store (7.9%). Similarly, the FinAccess household survey by the Central Bank of Kenya (CBK) finds that 59% of credit uptake has been from informal lenders with store credit being the most preferred source (29.7%) followed by family and friends (10.3%) (CBK 2019).

Perhaps the most alluring quality of informal credit is that they are more flexible compared with formal institutions which deal with loan repayment timelines and the accrual of interest more strictly (Atieno 2001; Diagne 1999; Johnson and Nino-Zarazua 2011). Johnson et al. (2010) suggest that informal saving groups afford women much-needed secrecy in their finances; this, Ojo (2022) argues, is particularly important in Africa where complicated gender relations persist. Wairire and Muiruri (2016) further suggest that the social embeddedness of informal financial groupings makes them pivotal to community development. Informal groups are also credited with spurring women entrepreneurship. A study by Richardson et al. (2004) that relied on surveys of small women-owned enterprises in Ethiopia, Tanzania, and Zambia found that significantly more women entrepreneurs turn to informal financing mechanisms to fund their businesses compared to men. Likewise, a study of table banking groups in Kenya by Gichuki et al. (2015) revealed that informal savings groups unlock credit for uncollateralized women which helps them expand their enterprises.

Having presented the argued benefits of informal financial services, we must also highlight ways in which informal financial services are perverse and by some accounts pernicious. Most of the arguments relate to the fact that these services are typically small in size with little room to scale up, lack in structure, lack basic consumer protections which make them generally exploitative, and tend to be expensive particularly as it relates to credit. For example, a Kenyan study by

Feather and Meme (2018) found that in some cases, the interest rate on loans from table banking groups can be as high as 20%. Even though this interest is recirculated into the group's pool, the high interest rates make loan repayment obligations harder to meet, leading to default. Many of the perceived or actual ills of informal financial services in the literature are summed up by Deborah James in her book *Money from Nothing* where she studies the provision of credit in post-apartheid South Africa. In the book, James (2014) provides a compelling account of how local money lenders, referred to as mashonisas, exploit poor black wage workers by offering them loans at exorbitant rates which lead to perpetual debt traps. She notes that mashonisas are borrowers themselves and that they borrow cheap and lend dear to those who do not qualify for formal credit, thus essentially "making money from money." She further describes the crude collection techniques employed by lenders including the confiscation of a borrower's debit card and identity document and the misuse of garnishee orders to recover debts directly from a borrower's meager wages.

Other drawbacks of informal financial services relate to the commitment and effort required to actively participate in informal savings and credit groups which Aldridge and Patterson (2002) point out as a key reason for the collapse of many of these groups. Tied to this is the time and effort taken to build social capital which is a requisite to access informal credit (Kilpatrick et al. 1999). The centerpiece of the widely popularized microcredit model was the notion of group lending which leveraged "informal insurance" from a borrower's network to effect loan repayments. This was seen as an effective way to bypass collateral requirements imposed by formal lenders, and some studies argue that it leads to much lower rates of loan defaults (Bouman 1995). Conversely, Anderson et al. (2009) posit that the lauded wisdom of group lending, specifically social sanctioning and its reputational effects, is not a solution unto itself and must be reinforced by external enforcement mechanisms.

The debate sketched out above begs a critical look at informal financial services to better understand why people turn to them amidst a rise in the provision of formal financial services, and more critically their contribution, if at all, to the broader mission of financial inclusion. In the next sections, we draw from two case studies – Zimbabwe and Kenya – where informal financial services play a considerably prominent role in the financial lives of people.

2.3 Informal Financial Services in a Time of Crisis: The Case of Zimbabwe

Over the last decade, there has been a spirited drive by formal financial institutions in Zimbabwe to reach out to the unbanked and underbanked populations, who mostly reside in the informal economy. The inception of the mobile money service EcoCash in 2011 by Zimbabwe's leading mobile network operator, Econet Wireless, held the promise of deeper financial inclusion for the millions of financially excluded Zimbabweans. The government, through the National Financial Inclusion Strategy (2016–2020), proposed a raft of interventions aimed at advancing financial inclusion in the country. This, according to the Zimbabwe National Financial Inclusion Strategy (2016–2020), included the lifting of barriers to financial inclusion for identified priority groups which included women, micro, small and medium entreprises (MSMEs), youth, and rural, and small-holder farmers (Reserve Bank of Zimbabwe 2016). In 2018, the government set up a deposit-taking MFI dedicated to women, the Zimbabwe Women's Microfinance Bank ostensibly to advance the financial inclusion of women and privilege them in the entrepreneurship arena. Commercial banks have also made some efforts to tailor their products and services to accommodate the financially excluded informal sector including the introduction of KYC-lite bank accounts and forging strategic partnerships. One such partnership is one between the Women's Bank and Knowledge Transfer Africa (KTA) which offer co-guaranteed micro loans to informal traders, mostly women.

Despite these efforts, the financial exclusion of women in Zimbabwe persists. There exist significant gender gaps in financial inclusion in Zimbabwe, with a gap of 10% in women's bank account ownership and a 7% gender gap in ownership of mobile money accounts (FinMarkTrust 2020). This is also illustrated in the provision of formal credit to women; between 2016 and 2020, women in Zimbabwe received less than 16% of all loans advanced by banks with that figure from 4% to 16% (ibid). These statistics depict a colossal failure in the financial inclusion of women in Zimbabwe by the formal financial system. This failure is deeply informed by broader economic challenges which have led to increased informal economic and financial activity as a way of cushioning people from the effects of failed formal structures.

2.3.1 A Financial Sector in Acute Crisis and the Resurgence of Informality in Zimbabwe

Zimbabwe has endured serious financial problems since the beginning of the twenty-first century which was mostly characterized by a prolonged period of hyperinflation (Nyamunda 2021) and (Mlambo 2014). In an attempt to counter rising inflation, the government through the Reserve Bank of Zimbabwe (RBZ) instituted a multi-currency system which eventually led to a defective and exploitative money exchange system according to Gukurume and Mahiya (2020) and Vasantkumar (2022). Further, Dzomira (2014) posits that Zimbabwe's financial sector has been increasingly fragile as evidenced by multiple bank closures with more than ten commercial banks becoming insolvent since 2004. To date, Zimbabwe is yet to rid itself of the aftershocks of the hyperinflation era (Ngwenya et al. 2018).

These conditions have significantly dampened financial inclusion efforts in Zimbabwe. Indeed, the RBZ acknowledges that the lack of confidence in the financial sector is one of the key constraints to financial inclusion in Zimbabwe. In particular, the deep economic crisis has cast significant doubts about the security and reliability of formal financial services and fomented conditions that made it nearly impossible to realize the benefits envisaged by mobile money. For example, Vasantkumar (2022) in a study of the chaotic currency situation in Zimbabwe pointed out that EcoCash agents (mobile money agents) take advantage of the rigged money exchange system to levy exorbitant fees on money transfers. He notes in addition that the RBZ severally "raided" people's bank accounts over the last 15 years, in some cases zeroing bank accounts and mandating arbitrary currency exchanges. This, coupled with the bank closures mentioned earlier, has significantly eroded Zimbabwean's confidence in their financial sector. Waning customer trust in the financial sector has in turn hampered financial inclusion as it has created a disincentive for the use of formal financial services.

The brief sketch of the state of affairs in Zimbabwe's financial sector paints a picture of a sector in deep crisis. This crisis has staged the ground for the thinking up of various informal and communal solutions as a way to help people deal with the effects of a failed formal financial sector. Of note is the resilience and ingenuity of the "people on the ground" to create parallel self-organized, self-service systems which are seemingly better suited to their needs. However, in the strict definition of financial inclusion as proffered by leading proponents of the discourse, this is not considered financial inclusion but rather simply a coping mechanism; we question this view.

The wider economic issues have led to the gradual shrinking of the formal economy in Zimbabwe and growth in the number of people depending on and joining the informal sector. Even though Jones (2010) notes that there has been a gradual increase in the number of men joining the informal sector, the ILO (2017) maintains that the informal sector in Zimbabwe is dominated by women. This gender skewness of the Zimbabwean informal economy can be traced from colonial history and a patriarchal system that instilled a "think vendor, think women" thinking (ILO 2017). Jones (2010) argues that the informal sector has shown to be

Zimbabwe's real economy since the turn of the twenty-first century. Despite contributing roughly 60% to gross domestic product (GDP), people operating micro and small businesses in the informal sector are overwhelmingly financially excluded. For example, a report on financial inclusion in Zimbabwe by FinMarkTrust (2020) points out that the informal sector received less than 5% of all banking sector loans and advances in 2017. This owes to their lack of physical assets to leverage as collateral for loans, and the lack of stable, verifiable cash flows among other factors. We infer, therefore, that there is an important connection between the financial exclusion of women and the informal economy in Zimbabwe.

Notably, the thriving informal sector has led to the devising of informal financing mechanisms which have brought services such as savings and credit closer to women. By organizing themselves in groups, women in the informal sector are able to save, get credit, and better manage their incomes (Jones 2010). As mentioned earlier, the archetype of informal financial services in Zimbabwe, as elsewhere in Africa, are saving groups also locally referred to as "kufusha mari" which are fashioned as ROSCAs, and "mikando" or "rounds" which are the local equivalent of ASCAs. These groups feature prominently in the informal economy in Zimbabwe. In a deliberate bid to help these groups be more sustainable, the Zimbabwe Chamber of the Informal Economy Association (ZCIEA) and other nongovernmental organizations (NGOs) play a role in training and building the group's capacities (Reserve Bank of Zimbabwe 2016).

Amid enduring national angst, and distrust of the formal financial system and the government more broadly, Zimbabweans have turned to informal financial services as a means to cushion themselves from the effects of a failed financial system. This is amplified by the fact that many people have been driven into the informal market where informal financial services are commonplace. Drawing on this, we argue that a fixation on the formal financial institutions to deliver financial inclusion in Zimbabwe is a false start; informal systems in the circumstances thus play a critical role in meeting people's disparate financial needs.

2.4 Charting Common Course: The Case of Chamas and Banks in Kenya

Informal savings and credit groups exist in a different set of circumstances from those of Zimbabwe discussed above. Kenya has had a fairly stable financial services sector post-independence, with the exception of blips in the 1980s and early 1990s when some relatively small financial institutions collapsed. Despite the steady growth of the sector after independence, banks were complacent in reaching out to poorer customer segments and sought to cater to wealthier segments, specifically people in formal employment, wealthier business people, and corporate clients. These conditions eventually led to the springing up of informal savings and credit groups after independence (Zeleza 1990). Some of these informal groups would later morph into full-fledged savings and credit cooperatives from the mid-1960s. Cooperatives were attractive to the financially excluded in that they catered to small depositors and offered loans that dispensed with the need for collateral. The microfinance wave in the 1990s and 2000s also sought to bank small depositors and advance credit using the group lending model.

The inflection point for banks came after the advent of the mobile money service, M-Pesa, in 2007, which irredeemably disrupted the financial services sector. M-Pesa significantly reduced the cost of money transfers, improved convenience, and later spawned other inventions such as digital payments and mobile lending services that rode on the rails of mobile money. Undoubtedly, a great part of M-Pesa's success story is rooted in the fact that the system has transcended formal settings and is widely used in informal settings where cash was prominently used in the previous dispensation.

Despite significant exploits in financial innovation that have put Kenya on the map, informal groups are still a common feature in Kenya, perhaps a testament to their often-overlooked contribution to the holistic financial inclusion of its members, most of whom are women. The 2021 FinAccess household survey found that 33.8% of the adult population in Kenya belonged to informal saving and credit groups, locally christened chamas (CBK 2021). The survey also noted that chamas are common in rural settings where financial exclusion is particularly exacerbated. Further, Kim (2022) and Johnson (2004) point out that chamas are more popular among women than men in Kenya.

In realization of the effectiveness of informal groups in mobilizing significant savings, commercial banks in Kenya have in the last decade been keen on tailoring a bouquet of financial products for chamas. Our analysis shows that savings and credit facilities for chamas are a standard offering by banks in Kenya, while some offer more specialized services. For example, the I&M Bank has a iMinVest chama account which they use to offer special insurance products for chama and individual members while the Kenya Commercial Bank (KCB) has fashioned investment services for informal savings groups through its Tuungane investment account. Likewise, Equity Bank also launched the Eazzychama platform which allows members to monitor their account activity to help the groups to help them manage their financial affairs more efficiently. This is evidently on the upward trend.

In addition to helping members of informal groups join the formal financial market, albeit indirectly, products and services tailored for informal groups help safeguard their savings. Thus preventing their "premature death," a key challenge facing chamas as noted by Johnson et al. (2010). Moreover, Beck and Demirgüç-Kunt (2008) point out that fashioning financial products for chamas has led to banks saving on back-end costs of mobilizing small deposits which have traditionally made serving the poor a low priority for banks. Johnson and Nino-Zarazua (2011) suggest that these small deposits would otherwise be directed to consumption needs instead of helping such households build their resilience against financial emergencies. This demonstrates a complementary relationship between formal and informal financial service providers, which is underplayed in the broader discursive space of financial inclusion.

Informality has long been presented in dominant modernization approaches as a continuum along which informal entities naturally progress toward formality. The Kenyan case described above argues otherwise; it depicts a situation where formal and informal financial service providers exist side by side and oftentimes complement each other. This complementarity proves or indicates that informal financial services are central to the financial inclusion of women in Africa.

2.5 Bringing It Together

This chapter spotlights the historical importance of informal financial services in African societies. The abstractions of solidarity or social movements are by-products of informal grassroots collectivism (Miller 2010). These high-minded ideologies motivated the formation of cooperative societies as an escape sting poverty and colonial oppression. In African society, the idealism of solidarity movements is enveloped in notions of Ubuntu (of southern African origin which loosely translates to oneness) and Harambee (Swahili, word meaning "pulling together") and are the motifs that run beneath informal financial services. We also observe that even though the concept of "informal insurance" was formalized and popularized in the microfinance years, it was not entirely new in African society but existed long before the 1980s. Our analysis suggests that attractive notions of collective action, independence, and autonomy, which formal financial services are to a great extent unable to offer, are the key reason why informal services have withstood the test of time.

We have also demonstrated how proponents of financial inclusion put a premium on formal financial services and the easy ways that they reduce the significance of informally provided financial services, which as we have highlighted are popular among women in Africa. Indeed, it is often surmised that "formal is best." This serves the interest of the elite at the expense of the poor who are subjects of financial inclusion initiatives; for corporations, this is in the form of increased returns while for the government this comes in the form of taxes. Formal institutions often exclude the poor from the formal financial system and even when new technologies and policies essentially force them to cater to this segment, Bateman et al. (2019) and Soederberg (2013) posit that the institutions frame the poor as a new profit frontier and use it as a basis for engaging in excessive risk-raking instead of making efforts to truly shore up vulnerable people. This, they argue, only serves to heighten socioeconomic vulnerabilities and is thus inconsistent with the broader vision of financial inclusion.

The emphasis on formal financial services takes the perspective of the "people on the ground" for granted, deprives them of their agency, and fails to consider that people know their circumstances best. Importantly, a perpetuation of this view risks the well-intentioned goal of financial inclusion being viewed as a forceful policy. We platform the view that the elitist obsession with formal financial services needs reconsidering especially in the African context if the benefits of financial inclusion are to be achieved and the gender gap in financial inclusion closed.

In the case of Zimbabwe, we discern that informal financial services are used (mostly in the informal sector) to counter the effects of a troubled financial sector. We infer from this case that there is a confluence between formal financial services, informal financial services, and informal economic activity which if exploited could deliver significant financial inclusion benefits. In the case of Kenya where the financial sector has been stable and is a leading market in mobile financial services, chamas which have a long history in the country still maintain their luster. Formal financial services recognize the remarkable intermediation role that informal groups play and have sought ways to "partner" with them. This is not to suggest that consumers pay no difference as to whether financial services are formally or informally provided. Indeed, people who have access to formal financial services find them useful for specific purposes and use informal financial services in certain circumstances. We argue that one can see a situation where informal financial services complement formal financial services and in dire situations can supplant formal financial services. Informal financial services, therefore, ought to be considered an important component of the financial inclusion project.

2.6 Conclusion

This chapter has sought to open up questions about the decoupling of financial services in an attempt to urge the reframing of financial inclusion discourse where proponents of financial inclusion, formal financial institutions, and people on the ground will be able to meet in reality. This, we argue, entails the recognition of informal financial services as mission critical and capacitating informal savings and credit groups. It also requires a reformulation of the relationship between formal and informal financial services where they can coexist and complement each other rather than attempts to formalize financial services provided informally.

References

Aldridge TJ, Patterson A (2002) LETS get real: constraints on the development of Local Exchange Trading Schemes. Area 34(4):370–381

Anderson S, Baland J-M, Moene KO (2009) Enforcement in informal saving groups. J Dev Econ 90(1):14–23

Atieno R (2001) Formal and informal institutions' lending policies and access to credit by small-scale enterprises in Kenya: an empirical assessment. (AERC Research Paper 111). African Economic Research Consortium. Retreived from http://erepository.uonbi.ac.ke/handle/11295/39700

Bateman M, Duvendack M, Loubere N (2019) Is fin-tech the new panacea for poverty alleviation and local development? Contesting Suri and Jack's M-Pesa findings published in Science. Rev Afr Polit Econ 46:1–16

Beck T, Demirgüç-Kunt A (2008) Access to finance: an unfinished agenda. World Bank Econ Rev 22(3):383–396

Bouman FJA (1995) Rotating and accumulating savings and credit associations: a development perspective. World Dev 23(3):371–384

CBK (2019) 2019 FinAccess Household survey. Central Bank of Kenya. Retrieved from https://

www.centralbank.go.ke/uploads/financial_inclusion/1035460079_2019%20FinAcces%20Report%20(web).pdf

CBK (2021) 2021 FinAccess Household survey. Central Bank of Kenya. Retrieved from https://www.centralbank.go.ke/uploads/cbk_annual_reports/322580908_2020%20Annual%20Report.pdf

Diagne A (1999) Determinants of household access to and participation in formal and informal credit markets in Malawi. International Food Policy Research Institute. Retreived from https://ebrary.ifpri.org/utils/getfile/collection/p15738coll2/id/125743/filename/125774.pdf

Dzomira S (2014) Analysis of bank failures during financial tumult in Africa-Zimbabwe: a historical review. J Gov Regul 3(3):75–80

Feather C, Meme CK (2018) Consolidating inclusive housing finance development in Africa: lessons from Kenyan savings and credit cooperatives. Afr Rev Econo Finance 10(1):82–107

FinMarkTrust. (2020) Zimbabwe financial inclusion refresh. In: Making access possible, UNCDF. FinMark Trust, Harare

Gichuki CN, Mutuku MM, Kinuthia LN (2015) Influence of participation in "table banking" on the size of women-owned micro and small enterprises in Kenya. J Enterp Communities 9:315

Gukurume S, Mahiya IT (2020) Mobile money and the (un) making of social relations in Chivi, Zimbabwe. J South Afr Stud 46(6):1203–1217

ILO (2017) Situational analysis of women in the informal economy in Zimbabwe. Edited by Labour and Economic Development Research Institute of Zimbabwe (LEDRIZ). International Labour Organisation, Harare

James D (2014) Money from nothing. In: *Money from Nothing*. Stanford University Press

Johnson S (2004) Gender norms in financial markets: evidence from Kenya. World Dev 32(8):1355–1374

Johnson S, Nino-Zarazua M (2011) Financial access and exclusion in Kenya and Uganda. J Dev Stud 47(3):475–496

Johnson S, Malkamäki M, Niño-Zarazua M (2010) The role of informal groups in financial markets: evidence from Kenya. Bath Papers in International Development and Wellbeing

Jones JL (2010) 'Nothing is straight in Zimbabwe': the rise of the Kukiya-kiya economy 2000–2008. J South Afr Stud 36(2):285–299

Kendall J, Voorhies R (2014) The mobile-finance revolution: how cell phones can spur development. Foreign Aff 93(2):9–13

Kilpatrick S, Bell R, Falk I (1999) The role of group learning in building social capital. J Vocat Educ Train 51(1):129–144

Kim K (2022) Assessing the impact of mobile money on improving the financial inclusion of Nairobi women. J Gend Stud 31(3):306–322

Miller E (2010) Solidarity economy: key concepts and issues. In: Solidarity economy I: building alternatives for people and planet, pp 25–41

Mlambo AS (2014) A history of Zimbabwe. Cambridge University Press

Morsy H, El-Shal A, Woldemichael A. (2019) Women self-selection out of the credit market in Africa.

Ngwenya B, Pelser T, Chivaura T (2018) Perceptions of post-multicurrency regime financial inclusion confidence challenges in Zimbabwe. S Afr J Econ Manag Sci 21(1):1–15

Nyamunda T (2021) 'Open for business' but Bankrupt: currencies, the 'new dispensation' and the Zimbabwean economy. J Asian Afr Stud 56(2):204–217

Ojo TA (2022) Digital financial inclusion for women in the fourth industrial revolution: a key towards achieving sustainable development goal 5. Afr Rev 14(1):98–123

Prahalad CK (2002) The fortune at the bottom of the pyramid. Strategy + Business 26:1

Prahalad CK, Hammond A (2002) Serving the world's poor, profitably. Harv Bus Rev 80(9):48–59

Reserve Bank of Zimbabwe (2016) Zimbabwe National Financial Inclusion Strategy (2016–2020) Harare. Reserve Bank of Zimbabwe, Zimbabwe

Richardson P, Howarth R, Finnegan G (2004) The challenges of growing small businesses: insights from women entrepreneurs in Africa. International Labour Office Geneva

Soederberg S (2013) Universalising financial inclusion and the securitisation of development. Third World Q 34(4):593–612

Vasantkumar C (2022) When the state tries to edit the dictionary… and fails: the return of the Zimbabwean dollar. Econ Soc 51(1):71–94

Wairire GG, Muiruri J (2016) Vyama in Kenya. In: The handbook of social work and social development in Africa, pp 292–303

World Bank (2019) Global ID coverage, barriers, and use by the numbers: an in-depth look at the 2017 ID4D-Findex. World Bank, Survey

Zeleza T (1990) The development of the cooperative movement in Kenya since independence. J Eastern Afr Res Dev 20:68–94

Zins A, Weill L (2016) The determinants of financial inclusion in Africa. Rev Dev Finance 6(1):46–57

3 Understanding Financial Inclusion in Africa from the Perspective of Saving and Borrowing Patterns

Queensley C. Chukwudum

Abstract

In Africa, women's access to finance is extremely low, leaving them at the mercy of systems that are not structured to favour them. The ability to save and borrow easily is critical to the financial empowerment of the African woman. This study, therefore, examines the savings and borrowing patterns of Africans through the gender and income status lenses. Although the general savings and borrowing distribution patterns of both genders are similar, a stark contrast is observed with respect to savings clubs. The findings indicate a strong inclination towards saving with a savings club for women compared to saving with a financial institution. Moreover, more women than men save (rather than borrow) to start, operate or expand a farm or a business even though the tails of the female borrowing behaviour showcases distinct characteristics in this regard. Additionally, it was established that men's saving and credit disposition significantly impacts women's financial behaviour. From an income status perspective, low-income countries are more visible in clusters that do not account for upper-middle-income countries. Therefore, to achieve the target for financial inclusion in Africa, it is recommended that policies which strengthen savings clubs should be enacted and thoroughly implemented whilst maintaining a well-established synergy between gender-targeted policies. Special finance-related quotas should be provided for low-income countries at regional and continental levels. All the highlighted recommendations have been aligned with selected United Nations Sustainable Development Goals.

Keywords

African women · Savings club · Cluster analysis · Linear regression · SDGs

3.1 Introduction

The growth of a nation is heavily dependent on the structure of its financial ecosystem which generally comprises public entities such as banks, regulatory frameworks, customer demands and financial technology. There is no doubt that these services thrive with higher usage. Thus, excluding any set of persons eligible to fully participate within this ecosystem depletes the nation's growth in the long run. The populations who are usually excluded in Africa are women and the elderly, according to Ozili (2020). Kodan and Chhikara (2013) pointed out the negative impact

Q. C. Chukwudum (✉)
Department of Insurance and Risk Management, University of Uyo, Uyo, Akwa Ibom State, Nigeria

T. A. Ojo (ed.), *Women and Finance in Africa*, Sustainable Development Goals Series,
https://doi.org/10.1007/978-3-031-53337-2_3

of financial exclusion, which includes unemployment, poverty, over-indebtedness and, subsequently, a life devoid of financial progress.

With the solid push to achieve the United Nations (UN) Sustainable Development Goals (SDGs) in particular, gender equality, poverty reduction and reduced inequality, financial inclusion has now been brought to the front burner. Shankar (2013) defined financial inclusion as the means of providing credit, insurance, payment service, and savings services to everyone who is qualified. This inclusion should cut across all barriers like demographic differences, income levels and gender, amongst others. Ozili (2020) rightly pointed out that governments and relevant stakeholders are beginning to embrace financial inclusion as a viable solution to economic empowerment and poverty alleviation. This, therefore, prompts the need to study financial inclusion from all possible perspectives.

The range of studies covering the financial inclusion subject abound in the literature. The main categories fall under studies that have examined the key determinants that are critical to a successful implementation of financial inclusion (Kumar 2013; Mhlanga 2021); those that have assessed the role of financial inclusion in promoting economic growth, development and financial stability (Kim et al. 2017; Mohan 2006; Hannig and Jansen 2010; Sarma and Pais 2011) and works that have investigated different ways to optimally achieve financial inclusion, for instance, through microfinancing, financial technology, re-framing and improving policies amongst others (Gabor and Brooks 2017; Donovan 2012; Marshall 2004; Arun 2021).

Upon the premise of re-framing and improving policies, many African women entrepreneurs and key stakeholders met for a workshop in Nairobi in 2016 to explore ways financially excluded women could be reached to enable them to participate and fully reap the benefits of financial inclusion (Abebe et al. 2017). One of the key takeaways is the need to identify and incorporate emerging and innovative mechanisms that can promote the uptake of financial services. Some of the innovative solutions highlighted include mobile money services, digital savings facilities, credit scoring–based social media and Google Analytics, flexible collateral, rotational savings and village banking schemes, amongst others. In other words, they tried to expand the traditional definition of financial inclusion to include emerging innovative methods and even informal ways to promote financial inclusion.

This chapter, therefore, seeks to understand financial inclusion from the quantitative perspective of the saving and borrowing patterns of women and men in Africa. This angle of thought has hardly been examined in the literature. Zins and Weill (2016) are the authors who have investigated the determinants of saving and credit behaviour in Africa by performing probit estimations on the 2014 World Bank's Global Findex database. They considered, in general, gender, age, income and education. The result of this study is compared with theirs since data from the same source is used. Country-wise, Helmy's (2015) study covered the saving behaviour of poor households in Egypt and the pathways towards accessing credit. The author's result indicates that households save more informally when there is clear access to credit. This is the more glaring case for women in Egypt. A similar study was conducted by (Lanidune 2021), who investigated village savings and loan associations in Ghana's west district as a model for sustainable financial inclusion in rural areas. The conceptual framework accounted for some key elements, such as the local realities, livelihood strategies and outcomes of poverty. There was a significant contribution of the existing savings and loan association towards the economic status of the locals. In East Africa, Lotto (2022) showed that households headed by men save more often for business than female-headed households. In the same vein, younger-head households save more than older-head households. Given the rise in digital financial services, Skogqvist (2019) studied the effect of mobile money on the savings behaviour of Kenyan citizens who have been financially excluded. Results from the study indicated that mobile money increases the propensity to save, especially within vulnerable rural groups. This revelation was also confirmed by Chamboko et al. (2018), who indicated that more African women than men receive and save money using mobile money. In general, though, more men in Africa use digital financial services than women.

No study, however, has examined the saving and credit behaviour within the context of the income status of African countries to the best of the author's knowledge, that is, low income, low-middle income and upper-middle income (UMI). This is another contribution to the literature that this chapter makes. Both informal and formal ways of saving and borrowing patterns are explored in this research to align with the need to explore all options as indicated by (Abebe et al. 2017). Such an investigation is needed to understand behavioural patterns that are either similar or different between genders, and countries based on the different income levels. This information can consequently be used to inform the formulation or re-framing of targeted gender-responsive and country-wise income status-sensitive policies.

Section 3.2 provides a brief overview of the data and the methodology applied. The empirical analysis is undertaken in Sect. 3.3 with the corresponding interpretations. Section 3.4 lays out the policy implications within the UN SDG framework. Conclusions are provided in Sect. 3.5.

3.2 Data and Methodology

The World Bank's Global Findex database has data from only three years (2011, 2014 and 2017). Given that 2011 has a lot of missing values for many African countries, only two years (Global Findex 2014 [released 2015] and Global Findex 2017[released 2018]) are used for this study (Demirguc-Kunt, et al., 2014; 2018). The data in the tables and figures has been gathered through surveys carried out in about 143 countries covering almost 150,000 persons globally. Thirty-nine African countries were extracted. It should be noted that the same countries have different years of data entry. For example, Nigeria has entries for the years 2014 and 2017. The analysis, therefore, does not focus on specific countries, instead it focuses on the yearly entries and puts all the countries under the African 'umbrella'. The level of financial inclusion is assessed based on many indicators. Questions ranging from the level of account penetration, the degree of financial services usage, the reasons and motivations for saving/borrowing and the informal finance alternatives are provided. The data is also disaggregated along the lines of gender, age, income and education. Table 3.1 displays the indicators/attributes that have been selected for this study, together with their assigned abbreviations.

The plot of each attribute related to saving and borrowing is depicted in Fig. 3.1.

From Fig. 3.1, different distributions can be observed. Some attributes like saved for farm/business male (sfm) display little frequency when compared to borrowing from family (bfamm) or borrowing for farm/business female (bff). Some of the histograms are more closely in the next section.

Simple methods are employed in this chapter. To avoid technical jargon, the methods are only be highlighted. The linear regression line is of the form:

$$E(Y) = E(a) + E(b)X \tag{3.1}$$

where X and Y are, respectively, the explanatory and the dependent variables. The slope of the line is b and the intercept is a (which is the value of y when $x = 0$). In clustering, the task is to group a set of objects such that the objects in the same group (called a cluster) share more similarities than to those in other groups or clusters. The cluster analysis requires one to specify in advance the number of clusters to extract.

3.3 Analysis and Results

3.3.1 Exploratory Analysis: Distributions of Selected Saving and Borrowing Attributes

In this section, the shape of the histogram is fundamental to explaining the characteristics of the data.

Table 3.1 Attributes relating to saving and borrowing

Saved to start, operate, or expand a farm or business, male (% age 15+)	Saved to start, operate, or expand a farm or business, female (% age 15+)	Saved for old age, male (% age 15+)	Saved for old age, female (% age 15+)	Saved at a financial institution, male (% age 15+)	Saved at a financial institution, female (% age 15+)	Saved using a savings club or a person outside the family, male (% age 15+)	Saved using a savings club or a person outside the family, female (% age 15+)	Saved any money in the past year, male (% age 15+)	Saved any money in the past year, female (% age 15+)
sfm	sff	som	sof	sfim	sfif	scm	scf	syrm	syrf

Borrowed for health or medical purposes, male (% age 15+)	Borrowed for health or medical purposes, female (% age 15+)	Borrowed to start, operate, or expand a farm or business, male (% age 15+)	Borrowed to start, operate, or expand a farm or business, female (% age 15+)	Borrowed from a financial institution, male (% age 15+)	Borrowed from a financial institution, female (% age 15+)	Borrowed from a financial institution or used a credit card, male (% age 15+)	Borrowed from a financial institution or used a credit card, female (% age 15+)	Borrowed from family or friends, male (% age 15+)	Borrowed from family or friends, female (% age 15+)	Borrowed from a savings club, male (% age 15+)	Borrowed from a savings club, female (% age 15+)	Borrowed any money in the past year, male (% age 15+)	Borrowed any money in the past year, female (% age 15+)
bmm	bmf	bfm	bff	bfin	bfif	bficm	bficf	bfamm	bfamf	bcm	bcf	byrm	byrf

Source: Author's compilation from Demirguc-Kunt, et al., (2015; 2018)

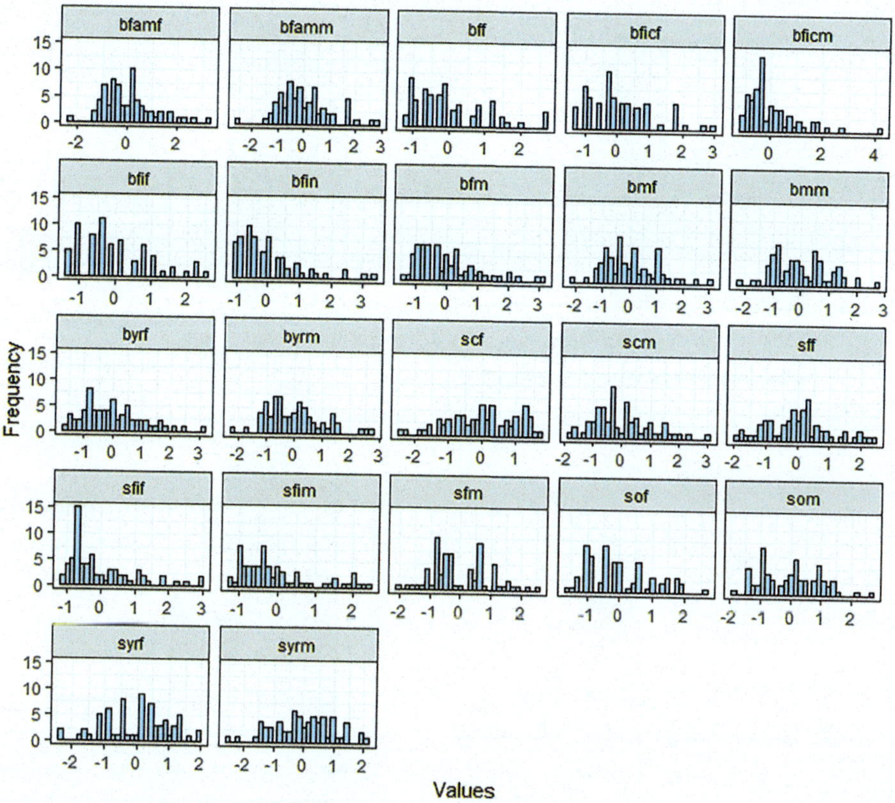

Fig. 3.1 Histograms of saving and borrowing attributes. (Source: Demirguc-Kunt, et al., (2015; 2018))

(i) Saving and borrowing for farm/business

While the distribution of females (in percentage) who save to start a farm/business is approximately normal, the distribution for those that borrowed is skewed to the right (Fig. 3.2a, b). This implies that as the percentage of women increases, the less they are likely to borrow to fund a farm/business. There is a high frequency of borrowing to start a farm/business for less than 10% of African women. As for saving for farm/business, the centre is close to 18%, implying that most of the women will fall within this range. Borrowing to start a farm/business for men (Fig. 3.2d) is similar to that of women – both are right skewed, but they differ significantly in the tails (extremes). Precisely, there is a smooth decline in the case of men but several jumps in the case of women. This interesting observation may suggest that as the percentage of women and men increase, the frequency of borrowing in both instances will reduce significantly. However, within this smaller frequency, women are more likely to borrow to start a farm/business than men are willing to.

So, differing observations can be noticed when it comes to funding a farm/business within the same gender for females in the case of saving and borrowing, and across different genders (between men and women) in the case of borrowing. This differs slightly from the findings of Zins and Weill (2016) whose result show that the probability of borrowing to start a farm/business decreases with being a woman, prompting the conclusion that men take up business-driven loans more. Here, they do not take into account the tail characteristics of the borrowing behaviour, which likely suggests that in the bulk of the distribution (which can represent the normal conditions), there is a higher probability that men will borrow, but in the tail (which can represent extreme/harsh conditions) the probability

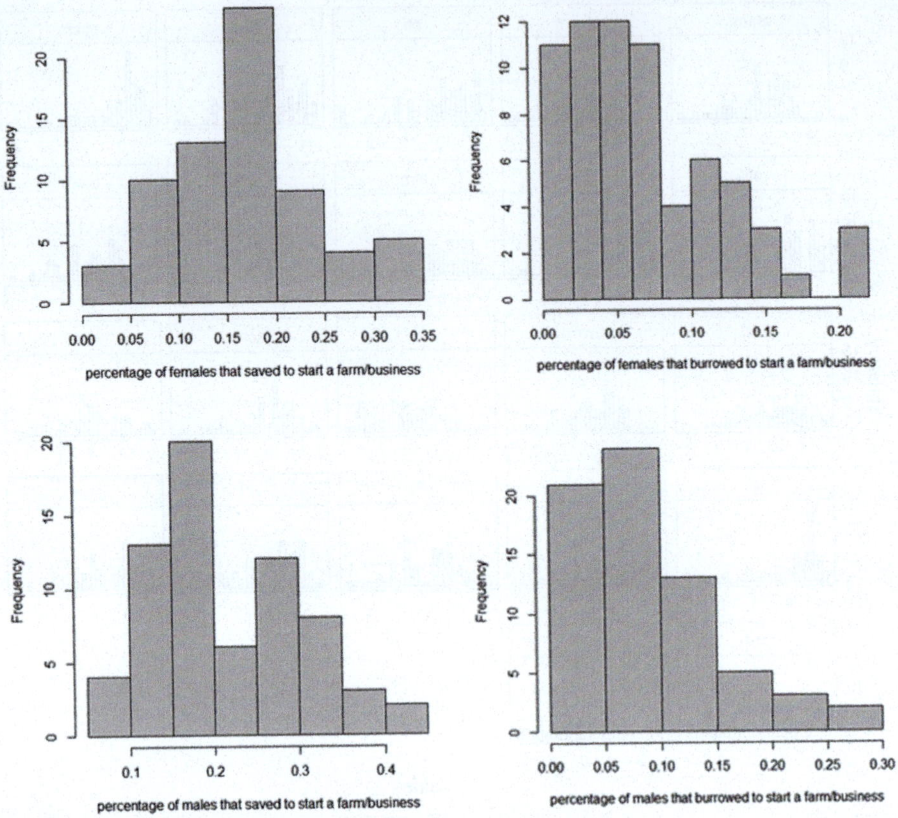

Fig. 3.2 Histogram for saving (**a** and **c**) and borrowing (**b** and **d**) for females (**a** and **b**) and males (**c** and **d**). (Source: Author's compilation from Demirguc-Kunt, et al., (2015; 2018))

switches in the favour of women. Nonetheless, this is still a preliminary result, hence the need for more analysis in the tails to ascertain this finding.

On the other hand, their result regarding the saving behaviour aligns with this study's findings – that is, saving is not significantly affected by gender (Fig. 3.2a, c). Nevertheless, a closer look at Fig. 3.2c indicates a histogram with two peaks. This suggests the possible presence of multiple model data. Thus, it may be more beneficial to disaggregate the data further to identify sub-populations of males who save to fund a farm/business.

(ii) Saving and borrowing from formal and informal sources

From Fig. 3.4, a glaring contrast in the way women in Africa save their money can be observed. While very few women save at a formal financial institution, the majority of women save using a savings club which is a more informal means of saving money. The left-skewed histogram depicts this in Fig. 3.4c which can be interpreted as follows: As the percentage of women increases, the greater will be the probability of women saving with a savings club. In essence, most women will seek for ways to make their savings within an informal rather than a for-

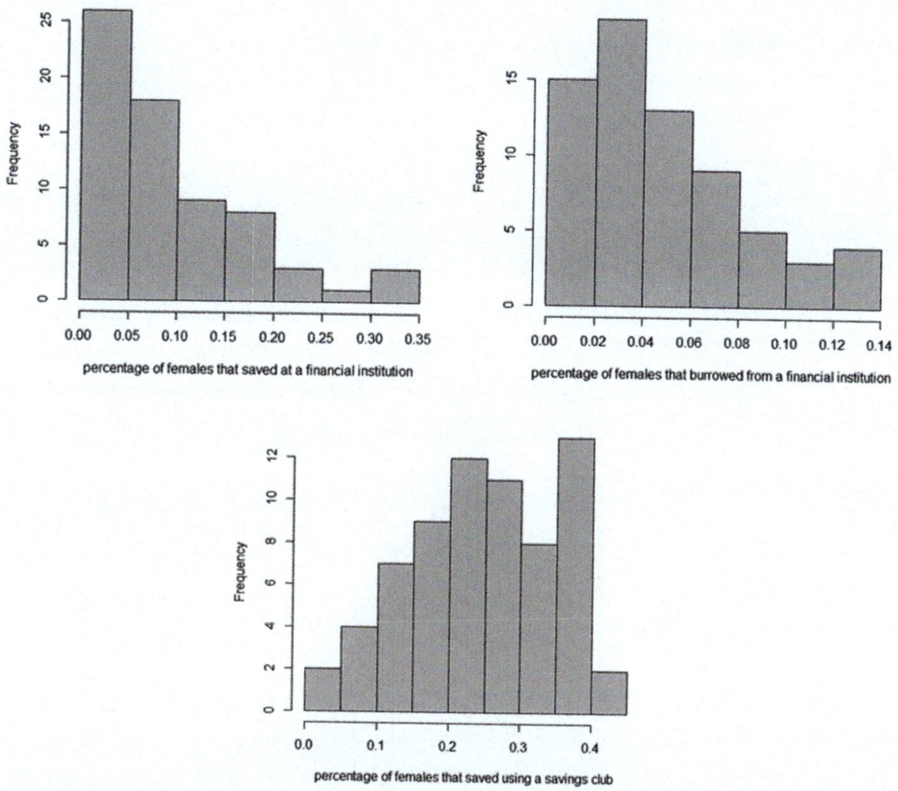

Fig. 3.3 Saving with and borrowing from a financial institution (**a** and **b**) and saving using a savings club (**c**) for females. (Source: Author's compilation from Demirguc-Kunt, et al., (2015; 2018))

mal setting. This finding aligns with Aterido et al. (2013), who also confirmed that African women are more likely to make use of informal services citing the presence of little to no discrimination as the reason. Quantitatively, Zins and Weill (2016) also confirmed that being a woman increases the likelihood of informal saving and decreases the probability of formal saving. The distribution of male saving behaviour follows a similar pattern regardless of whether it is a formal or informal set-up.

The bivariate plot in Fig. 3.5 generally suggests a positive linear relation between males and females concerning the following motivations – saving for farm/business, saving for old age and formal saving (Figs. 3.5a–c). In the case of savings club, the data points are more tilted towards the females (Fig. 3.5d). More associations can be observed from the correlation matrix (Fig. 3.5) which enables us to understand the relationship that exists between the savings and borrowing patterns of men and women.

3.3.2 Regression Analysis

From the correlation matrix, sff is moderately correlated with syrf and syrm. Further, the borrowing attributes correlated with the saving behaviour sfif are examined. Running a regression analysis to confirm further the associations gives the results as shown in Table 3.2.

From the regression analysis, syrm is a significant variable while syrf is not. This means that women are driven to save more to start a farm/business if a male relative has saved money within the past year. In a practical sense, Zins and Weill (2016) arrived at the same conclusion regarding the saving in the past year with respect to a woman. Their result indicated that the prob-

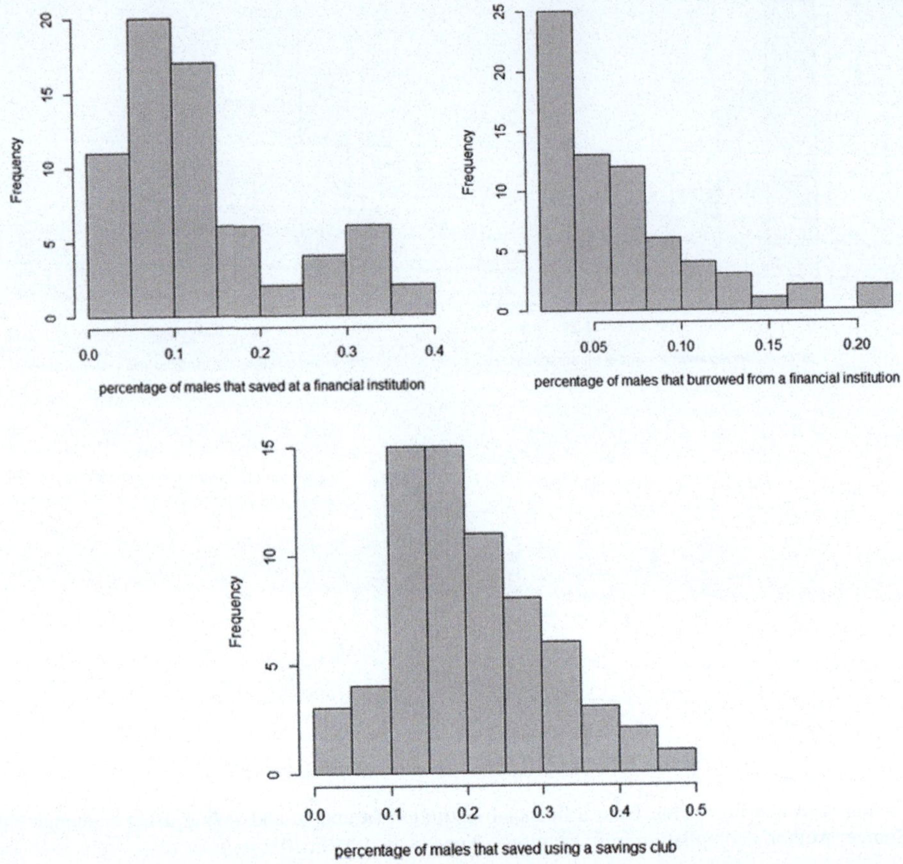

Fig. 3.4 Saving with and borrowing from a financial institution (**a** and **b**) and saving using a savings club (**c**) for males. (Source: Author's compilation from Demirguc-Kunt, et al., (2015; 2018))

ability of a woman setting aside money in the past 12 months is −2.1%, noting that women's reliance on informal finance does not help close the gender gap in formal finance. Relating their result to the findings here, syrf sure has no influence, but a new perspective emerges which has not been captured in previous studies. That is, if a man sets aside money in the last 1 year, a woman is more likely to save enough to start a farm/business. Possible explanations may include the following: (1) The woman is banking on the man's savings. This is the norm. (2) A husband's ability to save further motivates his wife to save more on her own. (3) A more practical scenario is that if a man sets aside money, he is better equipped to handle emergencies in the household. This reduces the financial, emotional and physical burden on the woman enabling her to save enough over time to buy a farm or start a business for herself.

For women who own a bank account, the significant borrowing attributes that positively affect the woman's ability to keep (and possibly increase) her savings at the financial institution include the formal credit given to men and the credit obtained by women who were given the option of borrowing directly or using a credit card. When the credit card option is left out, a negative association is observed – the woman is less likely to save. Thus, if the borrowing behaviour of men increases at a financial institution, women's savings behaviour will also increase. A similar conclusion was arrived at by Demirguc-Kunt et al. (2013) who noted, among other reasons, that a husband's adverse credit history was likely to increase his wife's exclusion from finan-

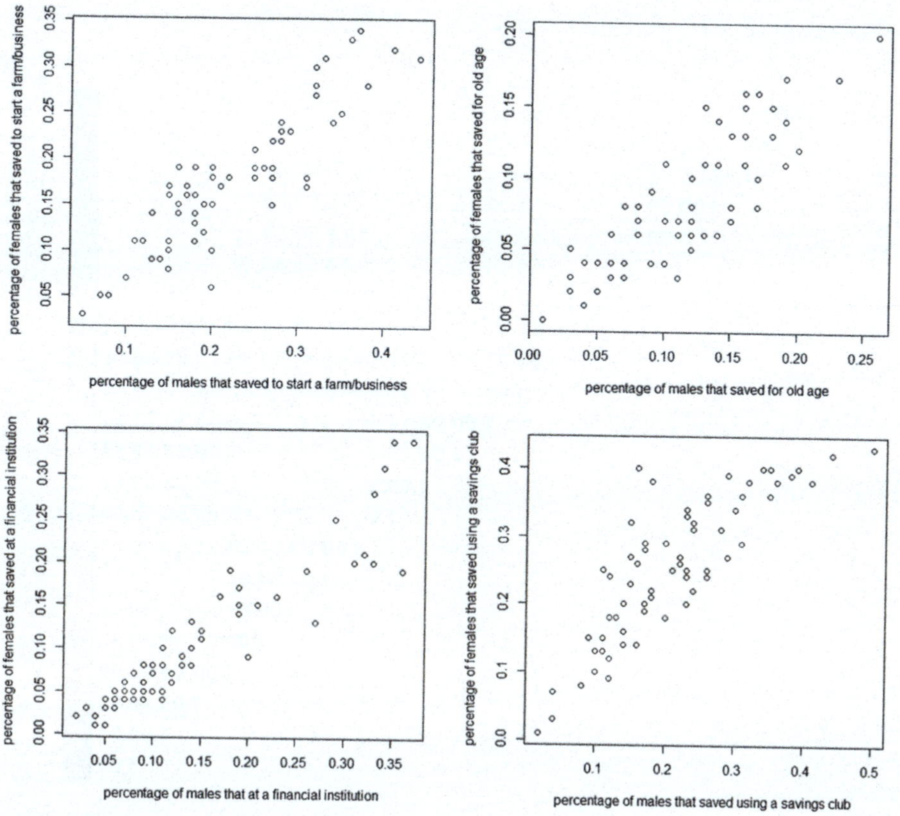

Fig. 3.5 Bivariate plots of male vs females based on specific attributes. (Source: Author's compilation from Demirguc-Kunt, et al., (2015; 2018))

cial formal services. This level of dependence needs to be drastically reduced over time to enable the financial choices of women to be made more independently.

3.3.3 Cluster Analysis

Combining all attributes (saving and borrowing), the cluster analysis was applied to identify distinct groups or financial behaviour across African countries based on their income status: low income, lower-middle income and upper-middle income. The k-means clustering is used to group the observations into clusters by testing different values of k. Figure 3.7 shows the different clusters for each k. To determine the optimal clusters, the elbow method is applied. A bend in the graph (Fig. 3.8) can suggest the appropriate number of clusters.

From the graph, the optimal cluster number is 4 (Fig. 3.7c). This gives the cluster sizes 7, 25, 13 and 23. Further, the proportion of total countries in each cluster with a specific status concerning spending and borrowing patterns is obtained, as shown in Table 3.3.

Using a similar classification as Zins and Weill (2016), if these four clusters are assumed to represent four quantiles, with the first to fourth quantile representing the poorest to richest groups, then cluster 2 will represent the poorest 25% since it has 18 low-income countries and no upper-income countries, cluster 3 will represent second 25%, cluster 4 the third 25% and cluster 1 will be the richest 25%.

For upper-middle-income (UMI) status, countries with a fairly developed financial system like South Africa and Mauritius (men and women) only appear in clusters 1 and 4 while other income level groups appear in all the clusters. Forty-three

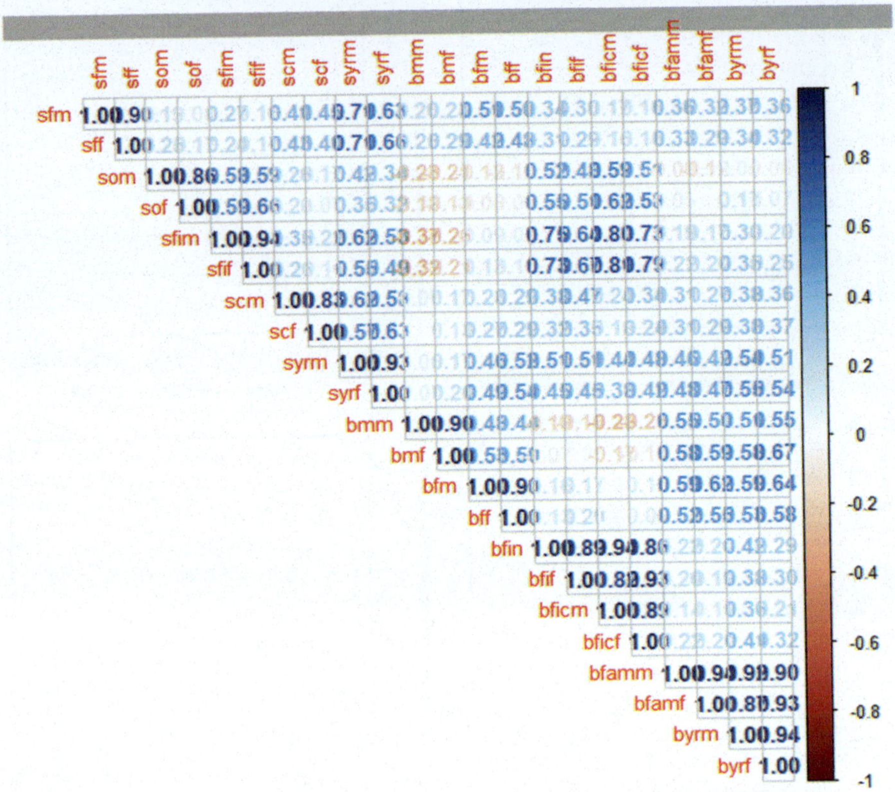

Fig. 3.6 Correlation matrix of saving and borrowing behaviour. (Source: Author's compilation from Demirguc-Kunt, et al., (2015; 2018))

Table 3.2 Estimates for the regression analysis

Variables	Estimates	Significance
Saving for farm/business (female)		
Intercept	−0.06512 (0.03038)	0.03582
Syrf	0.01365 (0.15657)	0.93080
Syrm	0.42025 (0.15064)	0.00691
Adjusted R-squared	0.482	
Saving at a financial institution (female)		
Intercept	−0.00657 (0.01248)	0.60034
Bff	0.18807 (0.11305)	0.10117
Bfin	0.82047 (0.26223)	0.00266
Bfif	−2.24117 (0.53179)	8.14e-05
Bfifc	2.41993 (0.37487)	1.76e-08
Adjusted R-squared	0.6992	

Source: Author's compilation from Demirguc-Kunt, et al., (2015; 2018)

percent of these African countries in cluster 1 are upper-middle earners, while the low- and lower-income countries share similar proportions of less than 30%. It can also be noted that in clusters where upper-middle-income earners do not showcase (2 and 3), low-income countries become highly dominant, probably implying that the presence of UMI earners highly constrains

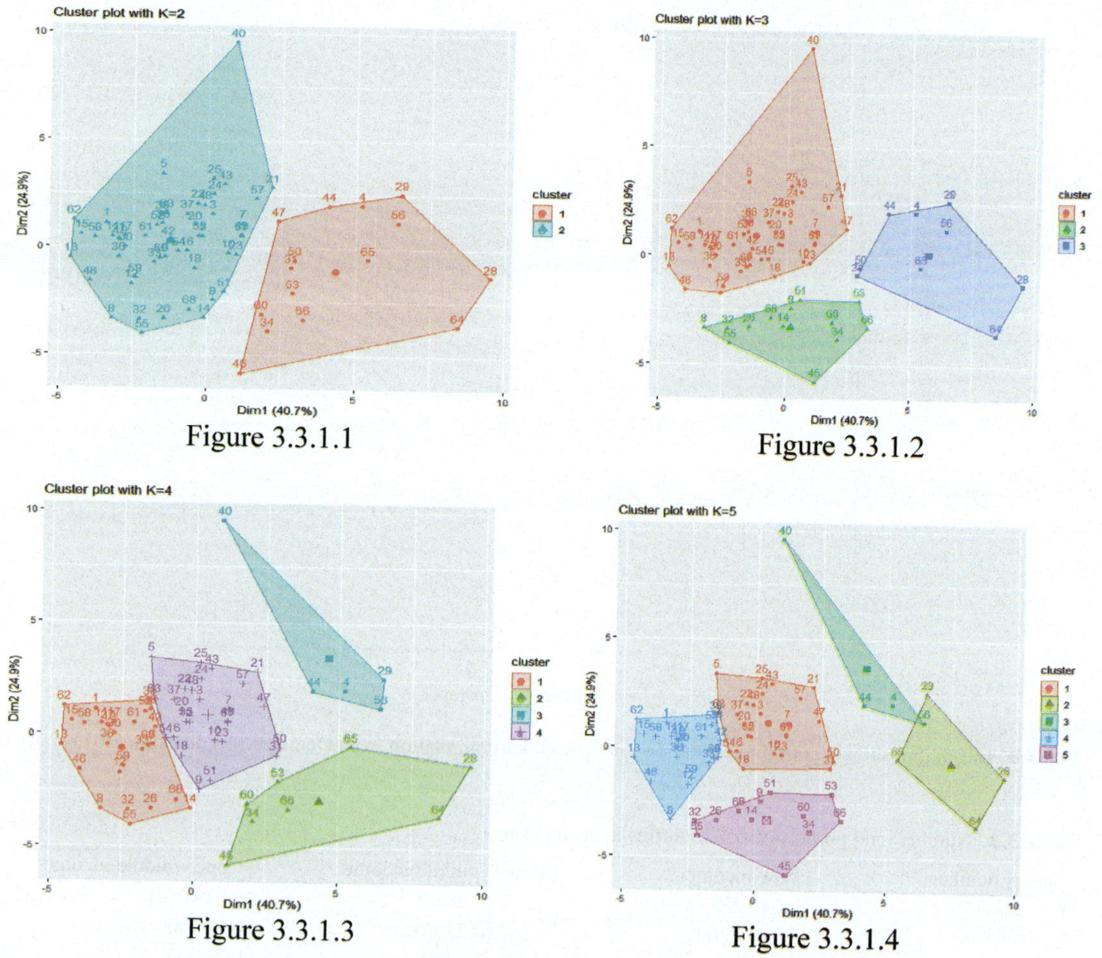

Fig. 3.7 Cluster plots with different k values. (Source: Author's compilation from Demirguc-Kunt, et al., (2015; 2018))

their opportunity to be financially included. A reason for this was highlighted by Thorsten and Cull (2013) who noted that credit from financial institutions is not evenly spread across the African population. This calls for regional stakeholders to design targeted policies for the different levels of income-earning countries because women in these countries (especially low income–earning countries) are likely to be even more marginalized financially. A similar strategy is sometimes applied when calls are made for scholarships or grants – the exclusion of some African countries that repeatedly appear typically gives room for less visible countries to earn a space.

3.4 Financial Inclusion, the SDGs and Implications for Policy

Thorsten et al. (2011) have pointed out the critical need for well-functioning and inclusive financial systems. They state that the poverty levels of countries with such systems tend to drop faster than those with weak and non-inclusive financial systems. In this section, five SDGs will be focused on, and their interactions with the saving and borrowing behaviours of Africans within a policy framework will be discussed.

1. To end poverty in all its forms everywhere (SDG 1), everyone needs to have access to financial services. Special consideration,

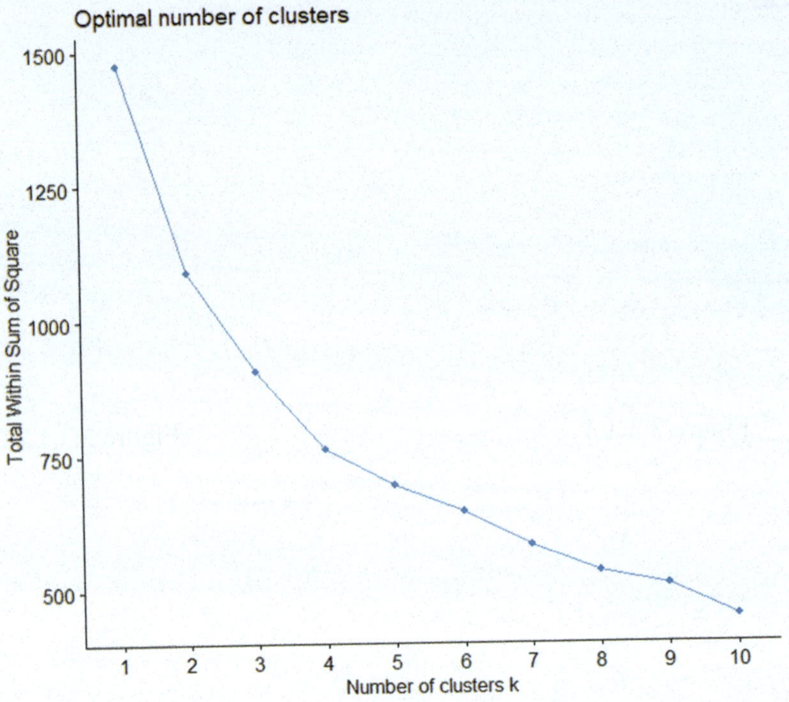

Fig. 3.8 Elbow plot showing optimal cluster. (Source: Author's compilation from Demirguc-Kunt, et al., (2015; 2018))

Table 3.3 Number (and proportion) of countries in each cluster

Cluster number	Low income	Lower middle income	Upper-middle income
1	2 (0.2857)	2 (0.2857)	3 (0.4286)
2	18 (0.7200)	7 (0.2800)	0 (0.000
3	10 (0.7692)	3 (0.2308)	0 (0.000
4	10 (0.4347)	7 (0.3043)	6 (0.2609)

Source: Author's compilation from Demirguc-Kunt, et al., (2015; 2018)

therefore, needs to be given to women who are more likely to save using savings clubs, as shown in Fig. 3.3c. This implies that government policies need to target savings groups, village banking schemes, and cooperatives, just to mention a few. These communities can potentially improve women's consumption and their ability to accumulate assets. Men's saving behaviour also significantly drives women's ability to save. This indicates the need for both parties in a given household to work together.

2. The need to promote sustainable agriculture (SDG 2), which is a panacea for food insecurity, brings to the fore the saving and borrowing behaviour relating to the purchase of farmlands. Since women prefer to save up to buy land, more opportunities should be created to enable them save efficiently. For instance, by providing adequate clean water facilities closer to home, women will have more spare time to earn more income as they will spend less time walking long distances in search of water. Basic financial literacy training programmes are also vital to promoting savings.
3. Ensuring healthy lives is the goal of SDG 3. It must be emphasized that building social net-

works is essential for mental well-being. If point (1) is catered to, the well-being of women will improve. Men will also benefit because they save with savings clubs (but to a lesser degree compared to women).

4. The disaggregated examination in this study has further confirmed that men and women do not always have the same characteristics. This perspective is key to promoting gender equality (SDG 5) because targeted policies regarding economic resources and access to finance can then be enacted to suit the characteristics that differ.
5. Reduced inequalities are SDG 10. The cluster analysis strongly suggests the need for special consideration along the lines of the income status of countries in Africa. Low- and lower-middle-income countries will reap more in terms of sustainable enterprise development if a graded or quota-share policy system is adopted with respect to the distribution of financial resources. This will help reduce the inequalities across the different levels of income-earning countries.

3.4.1 Prospects for Promoting Savings and Borrowing Patterns in Africa as Indicated by Other Authors

In line with improving the savings and borrowing patterns in Africa, Aryeetey and Udry (2000) recommended that several efficient 'localized' risk management strategies and households' activities be given enough attention within the formal financial markets as most financial savings on the continent are directed to the informal markets and institutions. Lotto (2022), on the other hand, strongly advocates for strengthening health and insurance systems as citizens' health and economic status are tightly intertwined. Another strategy employed by the Banking on Change and the Institute of Development Studies is to train youth savings groups. Six African countries (Egypt, Ghana, Kenya, Tanzania, Uganda and Zambia) were among the countries involved in the training programme covering the years 2013–2015. Flynn and Sumberg (2016) further examined the impact of the youth savings groups and the training they received after the programme was concluded. This was done for only Tanzania, Uganda, Zambia and Ghana. A gender-lensed data analysis approach has been suggested by Chamboko et al. (2018) towards designing targeted digital financial services that cater to the urgent needs of women. These analytics, as stated, should be supported with social network engagements.

3.5 Conclusion

This study has revealed that African men and women have differing saving and credit behaviour; these differences are key in designing and implementing financially inclusive gender-sensitive policies. However, the synergy between potential tailored policies must be well coordinated as men's financial behaviour significantly impacts women's financial behaviour. Additionally, targeted policies for the different income level countries are recommended given that low-income countries become more visible when upper-middle-income countries are not featured. A further examination of the tails of the saving for farm/business attribute is needed.

References

Abebe JO, Maina L, Ondiek J, Ogolla C (2017) Driving gender-responsive financial inclusion models in Africa. Background Paper, UN Women

Arun T (2021) Encouraging financial inclusion – improving policies and protecting families. https://www.essex.ac.uk/research/showcase/encouraging-financial-inclusion-improving-policies-and-protecting-families

Aryeetey E, Udry C (2000) Saving in Sub-Saharan Africa. Center for International Development, Working Paper No. 38, Harvard University

Aterido R, Beck T, Iacovone L (2013) Access to finance in sub-Saharan Africa: is there a gender gap? World Dev 47:102–120

Chamboko R, Heitmann S, Van Der Westhuizen M (2018) Women and digital financial services in sub-Saharan Africa: understanding the challenges and harnessing the opportunities. International finance Corporation, World Bank Group

Demirgüç-Kunt A, Klapper L, Singer D (2013) Financial inclusion and legal discrimination against women: evidence from developing countries, Policy Research Working Paper No. 6416. the World Bank, Washington, DC

Demirgüç-Kunt A, Klapper LF, Singer D, Van Oudheusden P (2015) The global findex database 2014: Measuring financial inclusion around the world. World Bank Policy Research Working Paper, (7255)

Demirguc-Kunt A, Klapper L, Singer D, Ansar S (2018) The Global Findex Database 2017: Measuring financial inclusion and the fintech revolution. World Bank Publications

Donovan K (2012) Mobile money for financial inclusion. Inf Commun Dev 61:61–73. https://doi.org/10.1596/9780821389911_ch04

Flynn J, Sumberg J (2016) Patterns of engagement with youth savings groups in four African countries. Institute of Development Studies, Research Report 81

Gabor D, Brooks S (2017) The digital revolution in financial inclusion: international development in the Fintech era. New Polit Econ 22:423–436. https://doi.org/10.1080/13563467.2017.1259298

Hannig A, Jansen S (2010) Financial inclusion and financial stability: current policy issues. Asian Development Bank Institute Working Paper, No. 259; 1–29

Helmy IAF (2015) Understanding the saving behavior of poor households: evidence from Egypt. MSc. Thesis, The American University in Cairo

Kim D-W, Jung-Suk Y, Kabir Hassan M (2017) Financial inclusion and economic growth in OIC countries. Res Int Bus Financ 43:1–14. https://doi.org/10.1016/j.ribaf.2017.07.178

Kodan AS, Chhikara KS (2013) A theoretical and quantitative analysis of financial inclusion and economic growth. Manag Labour Stud 38(1):103–133

Kumar N (2013) Financial inclusion and its determinants: evidence from India. J Financ Econ Policy 5:4–19. https://doi.org/10.1108/17576381311317754

Lanidune J (2021) Exploring financial inclusion with village savings and loan associations in Ghana's Wa-West District. PhD Dissertation Walden University

Lotto J (2022) Households' saving pattern and behaviour in East Africa. Cogent Bus Manag 9(1):2101418. https://doi.org/10.1080/23311975.2022.2101418

Marshall JN (2004) Financial institutions in disadvantaged areas: a comparative analysis of policies encouraging financial inclusion in Britain and the United States. Environ Plan A Econ Space 36:241–261. https://doi.org/10.1068/a3664

Mhlanga D (2021) Factors that matter for financial inclusion: evidence from sub-Saharan Africa – the Zimbabwe case. Acad J Interdiscip Stud 10(6):10.36941/ajis-2021-0152

Mohan R (2006) Economic growth, financial deepening, and financial inclusion. Address at the annual bankers' conference 2006, Hyderabad

Ozili PK (2020) Financial inclusion research around the world: a review. Forum Soc Econ 50(4):1–23. https://doi.org/10.1080/07360932.2020.1715238

Sarma M, Pais J (2011) Financial inclusion and development. J Int Dev 23:613–628. https://doi.org/10.1002/jid.1698

Shankar S (2013) Financial inclusion in India: do microfinance institutions address access barriers. ACRN J Entrep Perspect 2(1):60–74

Skogqvist JM (2019) The effect of mobile money on savings behaviors of the financially excluded. MSc. thesis, Södertörn University, Stockholm

Thorsten B, Cull R (2013) Banking in Africa. World Bank Policy Research Working Paper No. 6684. https://ssrn.com/abstract=2348417

Thorsten B, Maimbo SM, Faye I, Triki T (2011) Financing Africa: through the crisis and beyond. World Bank, Washington, DC

Zins A, Weill L (2016) The determinants of financial inclusion in Africa. Rev Dev Finance 6(1):46–57. https://doi.org/10.1016/j.rdf.2016.05.001

The Political Socioeconomic Fallout of the Mismanagement of COVID-19 Public Funds in South Africa

4

Michelle Beeslaar

Abstract

Gendered food security inequality elucidated the vulnerable position of women in South Africa who bear the brunt of the unresolved post-apartheid challenges of inequality, poverty and unemployment 28 years into the democratic era. The inability of women to access their constitutional right to sufficient food and water, in turn, limited opportunities for education, employment and entrepreneurship. The global political socioeconomic impact of the COVID-19 pandemic on women was worsened by the mismanagement of social response public funds. The study found that there is an adequate state policy framework coupled with enough public sector resources both budgetary and human. It acknowledged the negative and disproportionate impact of the pandemic on women noting this was due to the underlying reason for the remaining gender inequality being political in nature. This took the form of unsuitable and incapable senior cadre deployment, a rampant culture of corruption, lack of coordination, poor communication and a government structure not suited to optimal service delivery. State capture further paralysed commitments and initiatives. This study was conducted through a gendered conceptual lens with a systematic literature review methodology and thematic analysis of purposive data selection. Lessons noted were used to recommend locally innovated women-led agroecology food chain sector businesses proactively assisted by government implementation intervention and horizontal multi-actor governance. The study asserted the South African government's commitment to the United Nations General Assembly (UNGA)-adopted Sustainable Development Goal (SDG) 5 – Gender Equality – would not be met by the agreed target date of 2030. A turnaround of substantial progress was still feasible if SDG 2 – Zero Hunger – was prioritised with women's base physiological requirements met to be equitable participants in a financially inclusive and economically conducive space. The impediment was post-1994 politics, a stumbling block that needed to be mitigated.

Keywords

Mismanagement of social response · Public funds · Political socioeconomic impact · Covid-19 · Gendered food security

M. Beeslaar (✉)
Faolan Consulting, Pretoria, Gauteng, South Africa
e-mail: michellebeeslaar@faolanconsulting.com

T. A. Ojo (ed.), *Women and Finance in Africa*, Sustainable Development Goals Series,
https://doi.org/10.1007/978-3-031-53337-2_4

4.1 Introduction: COVID-19: A Political, Socioeconomic and Health Pandemic

4.1.1 A Health Response: South Africa (SA) and the United States (US)

COVID-19 is a mild to severe respiratory disease emanating from the novel coronavirus. First detected in Wuhan, China, in late 2019, its transmission is emitted through droplets or areoles from an infected person through coughing, sneezing and close-range verbal interaction. The disease from which most recover is more fatal in elderly individuals and those with comorbidities (World Health Organization 2021). The World Health Organization (WHO) declared it a pandemic in early 2020 with current active cases in 228 countries and territories, giving credence to the assertion of its unprecedented global political socioeconomic impact in the last century (WorldOMeter 2022). Effective 19 July 2022, there were 562,672,324 million confirmed infection cases worldwide with 6,367,793 million registered deaths (World Health Organization 2021). In its analysis of the African region, WHO Africa stated that to date, only one of every 71 cases had been reported. Deaths were anticipated to reduce to an average of 60 a day in 2022, a significant improvement considering COVID-19 ranked as the seventh leading cause of death after malaria on the continent in 2021 (World Health Organization Africa Region 2022).

The health response to COVID-19 highlights the fault lines of the anecdotally well-known diplomatic disparity between the Global North and South, the former colonisers, and its relationships today with previous colonies. More unexpectedly, it put a spotlight on the have and have-nots in some of the economically wealthiest and politically most powerful countries such as the US, brought to the fore by the #BlackLivesMatter momentum during the initial peak of worldwide infections. There is no correlation between vociferous protests in the wake of the publicly captured murder of an innocent Black man, George Floyd by a White police officer with increased infections during that period (Neyman and Dalsey 2020, 226). Still, the occurrence placed research focus on the disparities between African and Caucasian Americans with results signalling that even in counties where Black people are the minority, their infection and fatality rates are starkly higher, for example, both South Carolina and Chicago Illinois have a minority 30% Black population but constituted 51% and 70% of mortalities, respectively. A study attributes the consistent statistics of a minority populace being beset with the highest COVID-19 adverse impact due to a multifaceted set of socioeconomic conditions. These included crowded neighbourhoods coupled with cramped living spaces as Black families generally have to live intergenerationally and with multiple family members to afford rent or bond payments. They have a higher unemployment rate. During the height of COVID-19, they were employed in low-income but critical jobs, increasing exposure in spite of having low health insurance coverage. This exacerbated the higher levels of comorbidities without well-managed treatment including affordable access to healthy dietary options. All intensified if they were an African American woman (Peek et al. 2021, 286–292), eerily similar to Black SA women except they are the majority of the population.

The schism has sharpened as vaccines are being developed and made available. Vaccines were established to minimise individual, community and national health; additionally mitigating adverse economic impacts (GAVI 2022). A clear though unwritten diplomatic note verbale was sent to Africa that the lives of their Northern kin mattered more and would continue to do so whilst they dominated by holding the wealth to buy vaccines in such bulk that some expired. In several cases, these countries gave their citizens multiple doses irrespective of the WHO plea to have them distribute vaccines more evenly to provide an increased ability for overall geographic herd immunity. By the end of 2021, only 7.5% of more than 1 billion Africans had received their primary injection. The WHO vaccine director saw this as a flawed epidemiologic strategy as with the tendency of wealthy nations to donate vaccines seemingly socially responsible to poorer countries though it was often unused batches close to their expiry date (Nebehay and Mason 2021).

In a different approach, SA was one of the first and the largest bilateral recipients of the newly elected US President Joe Biden's manifestation of his first administration's commitment to donate 500 million vaccine doses globally. SA received 5.7 million Pfizer vaccine doses, sufficient to vaccinate 2.5 million South Africans though the population was set at over 58 million (U.S. Mission South Africa 2021; Statistics South Africa 2019). As of 29 July 2022, SA had 4,004,202 million confirmed cases of COVID-19 resulting in 101,977 thousand deaths. Vaccination uptake is at 37,025,629 million doses administered lowered not by the government's initial inability to facilitate the provision adequately but by high levels of vaccine hesitancy in the country, a country with an already poor civic education track record (World Health Organization 2022). A comprehensive literature review study published in the Expert Review of Vaccines suggested that vaccine hesitancy is a mirror to general SA societal issues of a lack of coordinated communication and transparency accompanied by public mistrust of state institutions (Cooper et al. 2021, 927).

4.2 COVID-19: A Political and Socioeconomic Pandemic

With over 200 countries and territories still having active COVID-19 cases, at their peak in 2020, it was inevitable that it would be compared to the twentieth-century phenomenon of the Great Depression which officially started in 1929 (WorldOMeter 2022). Similar to the COVID-19 pandemic, it had global economic repercussions with some countries more acutely affected such as the US where its countrywide shockwaves led to even the adoption of a new government economic approach. This Keynesian model lasted as the country's financial policy for almost four decades whereby the state should intervene to stabilise demand. This was in the face of a quarter of the population unemployed by 1933. Due to the outbreak of World War II from 1939 to 1945, recovery was a prolonged patchwork filled with hunger and homelessness (Smiley 2018). Different reasons for the debilitating economic downturn of the Great Depression rippled public effects, but they were comparatively as devastating especially for women then as they were at the pinnacle of the COVID-19 pandemic.

Globally, the pandemic had intersectional economic outcomes. Agricultural commodity prices declined by 20% in the US. A bilateral dispute between Russia and Saudi Arabia within the Organization of the Petroleum Exporting Countries (OPEC) caused, in one day, the biggest drop in the form of 24% off the price of Brent crude oil per barrel in three decades. Both effects were further detrimental to the chemical, food value chain and tourism industries in the US to the extent that with bonds at a ten-year low and income minimised, the government intervened with the Coronavirus Aid, Relief and Economic Security (CARES) Act (Nicola et al. 2020, 185–193). Trickling down, the socioeconomic dynamics were influenced by residential location and spatial layout, work and wage category and level of education (Lo et al. 2021). The same reasons that caused minorities in the US to be more exposed to contracting COVID-19 equally left them open to maximum negative consequences.

The estimation is that the global economic impact on the SA economy saw it contract by seven percent in 2020, resulting in potentially a nine percent increase in people living in extreme poverty (Statistics South Africa 2021; World Bank Group 2020). Even though SA is still considered an upper-middle-income country (UMIC), in the midst of the pandemic, it had a 2021 gross domestic product (GDP) of US 301.9 billion, but formal unemployment increased to 34.4% – a third of the population and not considering the damage of lockdowns to the informal sector (United Nations Population Fund 2019). Within this bleak SA economic climate, global factors and domestic management of COVID-19 caused many industries to lower their operational hours by an average of 26% to 40%. The least educated workers are the hardest hit. At a micro level, their households faced insufficient income for a full food basket increasing food insecurity (Arndt et al. 2020,111). SA as a developing economy places a significant burden on the informal economy to be the barrier between livelihoods and abject poverty (Khambule 2022). The government's almost laissez-faire administration of the informal economy provided a gaping lack of a social safety net during the pandemic. With these at-risk households locked into, by state security in many instances, one of the world's strictest lockdowns between 27 March and the end of July 2020, 47% of adults nationally surveyed in the Coronavirus Rapid Mobile Survey ran out of money to buy food for their household in April. Twenty-one percent of adults stated that in the week prior to the survey, an adult member of their household had gone hungry whilst 15% stated that a child had too (Mlaba 2020). The pressure of an unprotected informal economy ricocheted on especially women who were already the most vulnerable due to the uneven ratio of food insecurity, which is both an input and output factor of skewed gendered unemployment, underemployment, economically inactive and paid less for the same jobs as male counterparts or in general with jobs for women deemed less critical or easier and thus paid less (Casale and Shepherd 2021; van der Berg et al. 2022). This creates the conditions where one factor negatively reinforces the next and ends up placing women in a state where gender-based violence (GBV) can thrive as it embeds itself into a society where it has become a norm for, in 50% of cases, the victims to know and regularly share a household with their perpetrators (Statistics South Africa 2020).

4.3 COVID-19: South Africa's Gendered Political Commitments

The Universal Declaration of Human Rights (UDHR) was adopted in 1948 in Paris by the United Nations General Assembly (UNGA) in the aftermath of World War II. A war concluded with the devastation that ended with nuclear bombs. A war-weary diplomatic world crafted the UDHR that, though not legally binding in international law, placed responsibility on each nation to never repeat the human carnage by politically committing to a set of universal rights that were equally applicable to all human beings irrespective of nationality, race, religion, gender and other exclusionary criteria viewed as an input cause, allowing the extent of the war's destruction (Garrido et al. 2020). The UDHR is a key introductory point to SA's gendered political commitments because it formed a fundamental basis of many of this commitment documentation including the 1981 African Charter on Human and Peoples' Rights (ACHPR), 2015 Sustainable Development Goals (SDGs), Agenda 2063 and the 1996 democratic SA constitution.

The African Charter on Human and Peoples' Rights (ACHPR) was adopted by the Organisation of African Unity (OAU) in 1981 with acrimony but eventually ratified by all OAU member states in 1986 and incorporated wholly by the African Union (AU) in 1992 with 54 African states agreeing to domesticate supranational inherent rights, which akin to the UDHR included equality and economic development (African Union 2022). Almost three decades later, the AU's achievement of its blueprint for the African continent's success, Agenda 2063, is premised on the initial early phase development of youth and women. The Agenda specifically commits all member states to achieve gender equality by 2063 which entails an applied end to gender violence be it emotional, mental, physical or sexual abuse and discrimination across economics, politics and society. Positively, the pledge is made for the full financial inclusion of women to empower them to have equal access to the formal economy and thus place power in their hands to overcome the female poverty cycle (African Union 2015).

In 2015, the UNGA decided to chart a path towards achieving the earth's foundational basis for its future. The SDGs are 17 building blocks to forge a foundation for universal peace, prosperity and protection of our planet. The construction phase was designed for a whole-of-society approach that culminates in reaching the 17 goals by 2030 (United Nations 2022). This study's conceptual core is SDG 5 – Gender Equality (herewith referred to as SDG 5). The research detailed the causal socioeconomic links that produce society-wide gender equality. In turn, when SDG 5 is not addressed as an actionable policy imperative, the non-attainment or lack of progress results in not only the human rights of women being unmet but their human security threatened, hence the link in this work to SDG 2 – Zero Hunger (herewith referred to as SDG 2). SDG 5 recognised the improvement in the lives and livelihoods of millions of women in recent decades. This progress fell significantly short as women continued to be underrepresented in leadership positions across sectors, paid less for the same work, experienced far higher levels of violence and social discrimination and were overly employed in the more precarious informal sector. This diminished access to the formal economy and its socioeconomic benefits, namely access to affordable and sufficient nutritious food, safe water, quality primary healthcare, decent shelter and opportunities for education and skills-based training (United Nations 2022). SDG 2 was a statement that, in 2015, there were 790 million people across the globe who still grappled with hunger. The statement was solution orientated as it noted that for humans to actively participate in meeting the aims of all 17 SDGs, they needed to be food secure. Formally, the definition of food security used internationally is that it 'exists when all people, at all times, have physical, social and economic access to sufficient, safe and nutritious food which meets their dietary needs and food preferences for a healthy and active life' (United Nations 2006). The consequences of failing to urgently make substantive SDG 2 progress were all the linkages to SDG 5 broken down as women cannot self-actualise without their base physiological needs sustainably and sufficiently met (Maslow 1943, 430–437).

Fundamentally, Section 27 (1) (a), (b) and (c) of the SA Constitution accords the right to all who live in SA to access health care services including reproductive healthcare, sufficient food and water, and social services safety net where external circumstances reduce an individual's ability to provide a basic standard of living for themselves or their dependents (Constitution of South Africa 1996). Together, these documents provide an overwhelmingly strong legal environment and government directional maps to fulfilling both domestic and international commitments towards the human rights and human security of women in SA.

4.4 COVID-19: The Study's Aim and Research Methodology

Against the background of the COVID-19 pandemic, the research aim was to assess whether SDG 5 to be achieved significantly through SDG 2 progress was influenced by the corruption endemic to the SA if it carried through to the social response public funds and if so, the emanating results. The international and local data suggested that girls and women would experience the callous or careless maladministration most. The pandemic's early data indicated that gender equality and food security needed to become elevated policy priorities as they were human rights that had the proven potential to endanger human security and the stability of the state. With the impetus, it was key to provide credible research to keep the pedal on applied pressure to use the lessons learned for a better present and future for SA's girls and women.

Given the increasing pace of science and innovation within multidisciplinary research, care needs to be taken to not overlook the holistic tapestry of the particular research, particularly in the case of the COVID-19 crisis where policy and decision-makers wanted the science to make decisions in real time (Snyder 2019, 333–339). In practice, qualitative research with a systematic literature review and thematically collated information identified both the problem and allowed for an empirically based discussion on the key findings (Hawkins 2017, 1759–1760). Qualitative research was selected to describe the phenomenon of COVID-19, delve into the specific case of the mismanagement of public funds and recommend through thematic analysis means to improve the several challenges emphasised by the literature (Lichtman 2017, 86). The study stood on a gendered conceptual lens emanating from SDG 5 – Gender Equality.

4.5 Failed Governance and Systemic Government Corruption

Preceding the onset of COVID-19, SA remained amongst the most inequitable countries in the world with a Gini Coefficient of 0.63 (Statistics South Africa 2020; World Population Review 2022). COVID-19 did not cause the triple challenge of inequality, poverty and unemployment that the country has grappled with since the first democratic elections in 1994. In 2019, the economy grew by only 0.1%, and the private sector recorded the lowest profits in 14 years (World Bank 2021; Statistics South Africa 2021). When President Ramaphosa declared a State of Disaster in response to COVID-19 in March 2020, SA was hailed for its swift action and WHO-aligned response. Less than two weeks later, strict state security forces implemented some of the harshest lockdowns in the world at various levels for almost two years. The verifiable gains made by the ruling political party which had free and fairly won each election since 1994 were already being sullied by the state capture years under former President Zuma who is charged and awaiting trial for multiple alleged criminally corrupt deeds. The country's progress viewed through anecdotal and empirical cracks at the grassroots level 26 years since 1994 was already being questioned outside the onset of an unplanned pandemic with unequivocal statements that the government was failing to correctly prioritise the developmental state agenda (Moyo 2019).

The former Chairperson of the RSA Public Service Commission (PSC) recognised that whilst the pandemic placed a heavy burden on the state to deliver, this was a challenge long in the making (Sangwani 2020). It should have been resolved but as too much time lapsed and too few services were delivered or delivered to an insufficient number of people, it became more complex. The solution was the application of common sense. First, the constitution was the government's strategic plan for building a successful democracy. Second, committed and capable servant leadership were required to operationalise the constitution. Third, deployed and recruited public sector posts were to be filled by those post-specific qualified in education and experience to the level of expertise the job required. Fourth, the structure, operations and resource allocation had to be linked to the constitutional strategic plan. Fifth, cooperative and horizontal governance were needed whereby the three tiers of government – local, provincial and national – coupled with partnering across – academic, community, civil, religious and private – sectors broke the silo mentality and achieved mutually beneficial results in processes which were multi-actor, multidisciplinary and relatively power balanced and where resources were leveraged. Professor Sangwani looked post-pandemic to an active citizenry reclaiming their power to govern as envisioned in the Freedom Charter to apply pressure to make the hard decisions to resolve the challenges the public sector faces in the short term that will in the long term benefit the public in a living constitution.

Throughout the literature reviewed, three common issues arose consistently, namely the government's unabashed lack of accountability and transparency eroding the present low public trust in government and state institutions. The lack of Parliament's ability to fulfil its oversight function allowed the Executive – the President, Deputy President and Ministers in the Cabinet – to face minimal consequences for impaired decisions, undertakings made in Parliament not done and grave errors with public negative impact. It was suggested that Parliament instead rubber-stamped Executive decisions adopted long before placed on the parliamentary agenda and covered as best it could any allegations or formal investigation requests that may expose wrongdoing on the part of the Executive, individually or as a collective. This is argued to be how state capture was facilitated initially and allowed to grow to a mammoth extent with the aid of the ruling political party citing time immemorial that the party came before the government. Constitutional Chapter Nine institutions were weakened, under-resourced and without sharp enough teeth to hold the government to account as the courts could. The judiary option was an extremely expensive, requiring both excellent expertise and extended time. The government used this to its advantage to not implement stakeholder agreements, its own policy plans, international commitment and even provide basic constitutional rights inclusively and extensively. The lack of feasible opposition party alternatives, a weakened civil society operating in silos and citizens far removed from civic education of how the political system works and their own political power allowed for the political elite to be corrupted. This corruption grew to the magnitude of the entire state, with cadre deployment used as a ruling party reward system leading to disastrous outputs where these appointees were ill-equipped and out of their depth. This paved the way for a malaise amongst too many public servants who lacked direction, purpose in practice, oversight, training and mentoring which added to poor service delivery and the mismanagement of the public sector, contributing to the erosion instead of the renewal of the state (Fakir 2021). Without public service being a call of duty, the SA people, predominantly marginalised women, saw R34 of every R100 spent on the public sector budget allocation used for staff payments only, whereas if the resources were aligned to purpose, they could have achieved at minimum the foundation of a developmental state – food security and gender equality (Sangwani 2020).

The above can be summarised by the former and the current Presidents of SA. The current President, Cyril Ramaphosa, stated in 2018, 'We remain a highly unequal society, in which poverty and prosperity are still defined by race and gender' (Ramaphosa 2018). In 2018, he also noted that for the direly needed economic turn, 'Building growth, development and transformation, depends on a strong and capable state. It is critical that the size and structure of the state are optimally suited to meet the needs of the people and ensure the most efficient allocation of resources' (Ramaphosa 2018). Four years later, our own Stats SA produces data that we are spiralling downwards economically, and this is exacerbating our negative social behaviour. More recently, former President Thabo Mbeki called out his ruling party comrades including the current President warning of an SA 'Arab Spring' of people so hungry, homeless and hopeless that their best chance of survival lay in attempted revolution. He lambasted President Ramaphosa for failing to produce his National Action Plan within his promised 100 days from the delivery of his State of the Nation (SONA) address in February 2022. He called on the governing leadership to remember to tell the people the truth about the situation of the state for people to iteratively lead and work with the government in a renewed social compact to transparently resolve to not be a country in perpetual crisis before it is too late. Mbeki urged his fellow political party peers to carry out their duties as true revolutionaries who did not seek title or reward but found their purpose in serving others (Masuabi 2022). Four years have passed between 2018 and 2022, and whilst an Arab Spring event is unlikely at this stage given the current government and political trends, if a national coalition government is elected and formed in 2024, it should urgently address the building blocks for SA to reduce the current inequality, poverty and unemployment statistics in lived experiences by a significant margin, or there will be no human security without which there will be no state stability. In the next four years, the information contained in the literature was interpreted to render the best socioeconomic results focusing on women as the rock upon which to build this foundation, thus making SDG 5 and its linked SDG 2 two of the top ten priorities and resource allocation policy to practice megaprojects.

4.6 The Socioeconomic Impact of Mismanaging COVID-19 Public Funds: Hungry Women Cannot Achieve Financial Inclusion and Economic Development

The R500-billion stimulus package was always going to be vulnerable to corruption in a captured state at war with itself to free the cuckolds or to tighten the noose (Transparency International 2020). Private firms colluded with Cabinet

Ministers including the former Health Minister and senior public servants. The Special Investigating Unit (SIU) is probing over 75 companies for fraudulent activities with the Unemployment Insurance Fund responsible for special payments intended for employee's lockdown left without or with reduced income. The pandemic revealed how embedded corruption was in SA society and that it even penetrated medical science, whereby between April 2020 and June 2021, the total COVID-19 expenditure by government departments exceeded R138 billion (Abbasi 2020; Heywood 2021). Of this R14.8 billion, over ten percent is under investigation by the SIU with 214 cases already referred to the National Prosecuting Authority.

The knock-on effects of corruption stalling inclusive initiatives see women retain their financial risks and inability to manage the risks, be unable to pay for education, qualify for low-cost housing and access affordable, nutritious and safe food (Hendriks 2019). Overall, COVID-19 reiterated that SA faces a dichotomy where the country is nationally food secure, producing sufficient food in quantity and range that it is a net exporter of agricultural products, yet an estimated 6.8 million people are food insecure (Statistics South Africa 2019). There was something wrong when women and women-headed households, in particular, suffered COVID-19 lockdown hunger whilst a 'patriarchal capitalist system' made export profits throughout the food value chain. Gendered food security inequality is critical, with 58% of female-headed households living below the 2021 upper-bound poverty line of R1,335 (US 87.82); 39.2% of these households do not have an income with the probability to access one unlikely since 43.4% of females aged between 15 and 34 years are not employed or enrolled in education or skills training. A devastating outcome as SA emerges from the pandemic panic in a labour market already gender biased where women earn approximately 30% less, 60% of South African women-led households are now more reliant on social grants (Statistics South Africa 2020). The bottom line is even if effective measures were to be in place to include women in the agricultural value chain economy outside of minimum wage and/or seasonal work, the absence of meeting base physiological needs would render self-actualisation an incredibly uphill battle without state intervention (Maslow 1943, 430–437).

The mismanagement of COVID-19 public funds not only impacted SA women in terms of another failure of human rights but also moved the country squarely towards human security instability (Moyo 2019). The UNGA 2012 adopted resolution 66/290 which stipulated that state security now expanded to human security to address issues of socioeconomic justice, survival and dignity. This definition of security post-2012 went beyond war to those aspects that concern everyday human survival (Browning 2013). A month into hard lockdown, the *Daily Maverick* was already reporting that hunger would be the biggest cause of instability and threaten the country's ability to remain secure (Davis 2020). On the back of protests and looting, often with women and youth photographed in the forefront, the article used sources from academia, civil society, policy and political organisations to highlight six reasons why hunger posed the biggest COVID-19 human and therefore state security threat. First, the government department responsible for distributing the emergency food parcels is the South African Social Security Agency (SASSA), yet their offices were closed during the hard lockdown; they had the national capacity to distribute 300,000 parcels in April, yet research estimated by May that at least 3.2 million parcels would be required in Gauteng, one of the nine provinces. Second, the SA government has no singular national database to properly assess and administer socioeconomic support. Third, linked to this lack of a national database was the onerous bureaucracy placed on the most vulnerable to provide a series of documents which differed per province, was not well communicated and some not easily attainable with many government front offices closed, costing money and risk of breaking lockdown regulations, all to support their claim of vulnerability. Fourth, the allegations made by opposition political parties are regarding food parcel recipients having to be members of the governing party. The alleged corruption was followed up with a series of criminal charges laid at the South African Police Services (SAPS) including the capital city Tshwane branch of the South African National Civic Organisation (SANCO). Fifth, the need to stave off hunger fed into the transmission basis of the pandemic, placing the vulnerable at heightened risk of contracting COVID-19. Given their socioeconomic conditions, the contagious spread of the disease compounded in low to no-income areas. Sixth, SA individuals, non-profits, community organisations and businesses of all sizes did come forward to help, but there was a lack of central governance steering and coordination. This was most needed at the peak of lockdown where such parties, individuals or organisations potentially faced the wrath of both SAPS and the army which had been deployed in civilian areas under the National Disaster Management Act regulations for the slightest breach in regulations irrespective of the goodwill. The fear could reasonably outweigh the good intent given that by the end of May 2020, SAPS had arrested over 230 thousand people and killed 11 men all of whom were Black much as in the case with George Floyd mentioned above in the US (Haffajee 2020). A woman with children to care for or a woman wanting to help another in these conditions faced yet another avenue for possible abuse in a country where femicide is five times higher than anywhere worldwide and where SAPS reported in 2019/2020 that an average of 116 rapes occurred daily, supporting several studies that one in four SA women would be sexually assaulted in some form during their lifetime (African Health Organization 2021). The likelihood of them proactively seeking aid or being willing to enter into areas where it was most needed could plausibly be understood and lead to the assertion that the overall socioeconomic conditions women faced prior to COVID-19 were intensified during the pandemic lockdowns, worsening their lives and livelihoods as governance failed.

4.7 The Way Forward: Politics, Policy and Practice

COVID-19 wreaked political and economic damage globally, but it has also opened a window of opportunity to innovatively address issues of economic inequality. It empirically demonstrated that SDG 5 was not on track to be met in SA or universally as per the UNGA resolution. It also offered a glimpse into the equally dismal progress of meeting SDG 2. Beyond the disappointment, COVID-19 shone a light by linking the two and focusing on women's financial inclusion and economic development through the agricultural value chain whereby momentum on progress could be made.

Agriculture is the foundation of developing economies. SA, as one such economy, must therefore foster a healthy agricultural industry that contributes to the country's GDP, food security, social welfare and job creation through manufacturing and ecotourism while adding value to the country's wealth of raw materials (Moore 2021). The government committed to providing holistic support to start up, scale up and develop small, medium and micro enterprises (SMMEs) which bridged market gaps with agriculture, a key identified business area to empower urban and rural women (Department of Small Business Development 2021). Female-led agribusiness contributes to the domestic economy and opens female partnerships continentally whilst enabling women to proactively overcome their gendered food security inequality – workable gendered Pan African decolonial solutions in action.

In perspective, during the zenith of COVID-19, it was commodities that stabilised the economy (World Bank 2021). Agricultural exports comprised almost ten percent of the country's export earnings (International Trade Administration of the United States of America 2021). Smart agriculture uses co-production of knowledge and focuses on government resource allocation and market intervention assistance where needed, which is the route that makes the agroecological transition a practical and feasible solution for women gaining entry to the formal market. The study defines agroecology as a sustainable, socially just, scientific, locally designed approach by the communities who have intergenerational knowledge and are impacted by participants in the overall agricultural sector (Greenberg and Kleinbooi 2021a, b). Whilst government visible hand intervention often raises the ire of the neo-liberal free marketers, it is ironically supported in this case by an international legal instrument the International Code for the Marketing of Breastmilk Substitutes, which SA ratified (World Health Organization 2017). Balanced to the domestic reality of the 1.7 million households who experience hunger at some point in SA, agroecology leaves room for building livelihoods through the informal sector, making this an SA-locally responsive and inclusive approach (Statistics South Africa 2019). There are many plausible options that now need implementation. For example, investment in the digitisation of agribusiness is encouraged as more technologies can advance with the increased and strategic usage of cellular and fibre networks. In 2020, 99.8% of SA had access to 3G cell phones with 91.2% smartphones (Statista 2023).

This study established that SA has made political commitments to gender equality and food security both internationally and in apex law through the constitution. Government departments such as the Department of Small Business have pledged undertakings to transform the skewed economy that has women under-employed, paid significantly less, accessing minimal training and educational opportunities as well as increasingly falling below the poverty line and being dependent on social grants (Department of Small Business Development 2021; Statistics South Africa 2020). Furthermore, empirical research has uncovered a plethora of existing and waiting-to-be-developed local agro-ecological transition agribusinesses that within the ambit of the Fourth Industrial Revolution can be customised and co-produced for genuine financial inclusion of women starting in SA but with the real possibility to form a Pan African common market in this sector led by women – meeting SDG 5 for the African continent by African women using the agrarian knowledge of their ancestors to develop agricultural technology that is accessible and affordable for women to economically emancipate themselves with the aid of horizontal governance.

The policy exists. The practice is possible. The stumbling block to horizontal governance and indeed good governance is politics as illustrated by the state capture data. This is not insurmountable if civil society stops waiting and whining for a government to come to save them (Washington 2022). The Freedom Charter was adopted by the African National Congress in 1955 as it led the anti-apartheid movement. It reads till today as the vision for the quality of life for all South Africans from whence freedom was attained. The time has again come for the civil society that successfully mobilised awareness of the minority apartheid government's social injustice crimes against the humanity of the majority of South Africans to take courage from knowing that the power of society is in their hands to lead the change. For Africa, good governance does not require international monetary and donor organisation checklists. It requires the way the African people prior to their colonised conceptions of government and its institutions governed: people-led horizontal governance whereby the community made genuine input into the decisions and oversight of how they were implemented. A continent capable of leading the knowledge economy is a continent that can modernise its own conception of good governance and apply it – universally as monitored by the people.

4.8 Conclusion

The ending to this chapter lies in a start to a realistic post-pandemic phoenix rising by South Africans to collaboratively work together to place impetus and urgency on achieving SDG 5 – Gender Equality linked to SDG 2 – Zero Hunger. At the launch of the COVID-19 Global Humanitarian Response Plan hosted by the United Nations Department of Economic and Social Affairs which houses the SDGs, the way forward was clear. Yes, COVID-19 impacted the most vulnerable, especially women, children and the elderly falling into the lowest economic brackets disproportionately. No, not enough was being done, and countries globally needed to partner now more than ever in modern history because the pandemic was not simply a worldwide challenge to health but to societal socioeconomics (United Nations Department of Economic and Social Affairs 2020). This remains the opportune moment for taking the lessons of COVID-19 and turning the tide against the SDGs being about geopolitical agendas or ever-needy and poorly governed countries, about handouts from the colonials to the colonised and the widening gap between the North and South like a US civil war of old. Politics, policy and practice aside, the way ahead in SA and globally is sustainable, developmental and gender-equitable societies – SDGs – that is simply #HumanityFirst.

Acknowledgements My gratitude for the guidance, support, partnerships, mentoring and affiliation with some of the organisations are extended in alphabetical order:

- British Academy funded Writing and Research in Inequality in Africa (WARIA)
- Department of Political Sciences and International Relations, Faculty of Humanities at the University of Pretoria
- Morne Beeslaar and Faolan Consulting

References

Abbasi K (2020) Covid-19: politicisation, 'corruption,' and suppression of science. BMJ 371(November):m4425. https://doi.org/10.1136/bmj.m4425

Africa Health Organisation (2021) Gender-based violence fact sheet: South Africa. 14 April 2021. https://aho.org/news/gender-based-violence-fact-sheet-south-africa/

African Union African Charter on Human Rights (2022) African charter on human rights. African Commission on Human and Peoples' Rights 2022. https://achpr.au.int/en/about/history

Arndt C, Davies R, Gabriel S, Harris L, Makrelov K, Modise B, Robinson S, Simbanegavi W, Van Seventer D, Anderson L (2020) Impact of Covid-19 on the South African economy an initial analysis SA-TIED. Working Paper 111. https://www.ifpri.org/publication/impact-covid-19-south-african-economy-initial-analysis

Browning CS (2013) International security. A very short introduction. Oxford University Press, New York. https://www.veryshortintroductions.com/view/10.1093/actrade/9780199668533.001.0001/actrade-9780199668533

Casale D, Shepherd D (2021) The gendered effects of the Covid-19 crisis and ongoing lockdown in South Africa: evidence from NIDS-CRAM waves 1–3, vol 39, p 644

Cooper S, van Rooyen H, Wiysonge CS (2021) COVID-19 vaccine hesitancy in South Africa: how can we maximize uptake of COVID-19 vaccines? Expert Rev Vaccines 20(8):1–13. https://doi.org/10.1080/14760584.2021.1949291

Davis R (2020) The biggest lockdown threat: hunger, hunger, everywhere. Daily Maverick, April 17, 2020. https://www.dailymaverick.co.za/article/2020-04-17-the-biggest-lockdown-threat-hunger-hunger--everywhere/

Fakir E (2021) Ebrahim Fakir I Parliament's stature and standing on the precipice of withering away. News24. August 21, 2021. https://www.news24.com/news24/opinions/analysis/ebrahim-fakir-parliaments-stature-and-standing-on-the-precipice-of-withering-away-20210821

Garrido VM, Mthembu P, Wilkins A (eds) (2020) The global politics of human rights: bringing the universal declaration of human rights into the 21st century. The Berlin Forum on Politics, Institute for Global Dialogue, RECLAIM! Universal Human Rights Initiative and the University of South Africa, Berlin/Pretoria. https://www.africaportal.org/publications/global-politics-human-rights-bringing-universal-declaration-human-rights-udhr-21st-century/

GAVI (2022) Vaccines work. Www.gavi.org. https://www.gavi.org/vaccineswork

Greenberg S, Kleinbooi K (2021a) Minutes of the Food Governance Community of Practice Workshop on Agroecological Transitions and Local Governance

Greenberg S, Kleinbooi K (2021b) Agroecological transitions and local governance. Minutes of the Centre of Excellence in Food Security Food Governance Community of Practice Workshop. November 17, 2021

Haffajee F (2020) Ramaphosa calls 11 lockdown deaths and 230,000 arrests an act of over-enthusiasm-really!. Daily Maverick, June 1, 2020. https://www.dailymaverick.co.za/article/2020-06-01-ramaphosa-calls-11-lockdown-deaths-and-230000-arrests-an-act-of-over-enthusiasm-really/

Hawkins MJ (2017) Thematic analysis. In: Allen M (ed) The SAGE Encyclopaedia of communication research methods. Sage, Thousand Oakes, pp 1767–1760. https://doi.org/10.4135/9781483381411

Hendriks S (2019) The role of financial inclusion in driving women's economic empowerment. Dev Pract 29(8):1029–1038. https://www.tandfonline.com/doi/full/10.1080/09614524.2019.1660308

Heywood M (2021) Covid-19 corruption tops R14-billion but to bust criminals we need to drastically boost prosecution services and courts. Daily Maverick – Maverick Citizen. September 21, 2021. https://www.dailymaverick.co.za/

article/2021-09-21-covid-19-corruption-tops-r14-billion-but-to-bust-criminals-we-need-to-drastically-boost-prosecution-services-and-courts/
International Trade Administration of the United States of America (2021) South Africa country commercial guide: agricultural overview. https://www.trade.gov/country-commercial-guides/south-africa-agricultural-sector
Khambule I (2022) COVID-19 and the informal economy in a small-town in South Africa: governance implications in the post-COVID era. Edited by Sandro Serpa. Cogent Soc Sci 8(1). https://doi.org/10.1080/23311886.2022.2078528
Lichtman M (2017) Conceptualising research approaches. Sage, London, pp 81–96. https://doi.org/10.4135/9781544307756
Lo C-H, Nguyen LH, Drew DA, Warner ET, Joshi AD, Graham MS, Anyane-Yeboa A et al (2021) Race, ethnicity, community-level socioeconomic factors, and risk of COVID-19 in the United States and the United Kingdom. EClinicalMedicine 38(August):101029. https://doi.org/10.1016/j.eclinm.2021.101029
Maslow A (1943) A theory of human motivation. Psychol Rev 50(4):430–437
Masuabi Q (2022) Thabo Mbeki slams ANC for failing on unemployment, poverty, inequality. Daily Maverick July 22, 2022. https://www.dailymaverick.co.za/article/2022-07-22-thabo-mbeki-slams-anc-for-failing-on-unemployment-poverty-inequality/
Mlaba K (2020) 3 things that have increased food insecurity in South Africa this year. Global Citizen October 27, 2020. https://www.globalcitizen.org/en/content/issues-increase-food-insecurity-south-africa-covid/
Moore M (2021) SA's agriculture sector – a key driver of economic growth. Future Growth Asset Management, South Africa. https://www.futuregrowth.co.za/insights/sa-s-agriculture-sector-a-key-driver-of-economic-growth/
Moyo B (2019) The right to food in South Africa: we need a manifesto for food justice. The Daily Maverick. https://www.dailymaverick.co.za/article/2019-03-05-the-right-to-food-in-south-africa-we-need-a--manifesto-for-food-justice/
Nebehay S, Mason J (2021) WHO warns against vaccine hoarding as poorer countries go without. Reuters, December 9, 2021, sec. Healthcare & Pharmaceuticals. https://www.reuters.com/business/healthcare-pharmaceuticals/who-warns-against-vaccine-hoarding-poor-countries-go-without-2021-12-09/
Neyman G, Dalsey W (2020) Black lives matter protests and COVID-19 cases: relationship in two databases. J Public Health 43(2):225. https://doi.org/10.1093/pubmed/fdaa212
Nicola M, Alsafi Z, Sohrabi C, Kerwan A, Al-Jabir A, Iosifidis C, Agha M, Agha R (2020) The socio-economic implications of the coronavirus and COVID-19 pandemic: a review. Int J Surg 78(1):185–193. https://doi.org/10.1016/j.ijsu.2020.04.018
Peek M, Simons RA, Parker WF, Ansell DA, Rogers SO, Edmonds BT (2021) COVID-19 among African Americans: an action plan for mitigating disparities. Am J Public Health 111(2):286–292. https://doi.org/10.2105/ajph.2020.305990
Ramaphosa C (2018) 2018 State of the Nation Address. Presented at the Opening of Parliament, February 16
Sangwani S (2020) Pandemics and their impact on public sector performance: how to build the envisioned capable state. The National School of Government Workshop Minutes June 25, 2020
Smiley G (2018) Great depression. Econlib https://www.econlib.org/library/Enc/GreatDepression.html
Snyder H (2019) Literature review as a research methodology: an overview and guidelines. J Bus Res 104:333–339. ISSN 0148-2963. https://doi.org/10.1016/j.jbusres.2019.07.039
South Africa. Department of Small Business (2021) Small business development and united nations development programme launch study on the impact of COVID-19 on micro and informal businesses in South Africa. Department of Small Business Media Release. https://www.gov.za/speeches/small-business-development-and-united-nations-development-programme-launch-study-impact
South Africa. Stats SA (2019) Towards measuring the extent of food security in South Africa: an examination of hunger and food inadequacy. Government Printing Works, Pretoria. http://www.statssa.gov.za/?p=12135
South Africa. Stats SA (2020) How unequal is South Africa?. http://www.statssa.gov.za/?p=12930
South Africa. Stats SA (2021) GDP quantifying SA's economic performance in 2020. March 9, 2021. https://www.statssa.gov.za/?p=14074
South Africa. The Constitution of the Republic of South Africa (1996). https://www.gov.za/documents/constitution-republic-south-africa-1996
Statista (2023) Smartphone users in South Africa 2014–2023 | Statista. Statista Statista January 18, 2023. https://www.statista.com/statistics/488376/forecast-of-smartphone-users-in-south-africa/
Transparency International (2020) In South Africa, COVID-19 has exposed greed and spurred long-needed action against corruption. Corruption Watch September 4, 2020. https://www.transparency.org/en/blog/in-south-africa-covid-19-has-exposed-greed-and-spurred-long-needed-action-against-corruption
U. S. Mission South Africa (2021) United States Donates 5.7 Million COVID-19 Doses to South Africa. U.S. Embassy & Consulates in South Africa. https://za.usembassy.gov/united-states-donates-5-7-million-covid-19-doses-to-south-africa/#:~:text=the%20.gov%20website.- Accessed 28 July 2021
United Nations (2022) Gender equality and women's empowerment. United Nations Sustainable Development Goals. https://www.un.org/sustainabledevelopment/gender-equality/
United Nations Population Fund (2019) Republic of South Africa facts and figures: Sexual reproductive rights.

https://www.unfpa.org/data/transparency-portal/unfpa-south-africa
United Nations. Department of Economic and Social Affairs (2020) Everyone included: social impact of COVID-19. March 23, 2020
United Nations. Food and Agriculture Organisation. Food security. Policy Brief, 2. June, 2006. https://www.fao.org/fileadmin/templates/faoitaly/documents/pdf/pdf_Food_Security_Cocept_Note.pdf
van der Berg S, Patel L, Bridgman G (2022) Food insecurity in South Africa: evidence from NIDS-CRAM wave 5. Development Southern Africa:1–16. https://doi.org/10.1080/0376835x.2022.2062299
Washington K (2022) Minutes of the workshop on urban women-led agribusiness in South Africa. US Spaces in South Africa June 1, 2022
World Bank (2020) Sub-Saharan Africa. South Africa. Poverty & Equity Brief April 2020. https://databank.worldbank.org/data/download/poverty/33EF03BB-9722-4AE2-ABC7-AA2972D68AFE/Global_POVEQ_ZAF.pdf
World Bank Overview (2021). https://www.worldbank.org/en/country/southafrica/overview#1
World Health Organization (2017) International code for the marketing of breast-milk substitutes. Frequently Asked Questions. 2017 Update. 2017. https://apps.who.int/iris/bitstream/handle/10665/254911/WHO-NMH-NHD-17.1-eng.pdf
World Health Organization (2021) Coronavirus disease (COVID-19). World Health Organization. https://www.who.int/health-topics/coronavirus#tab=tab_1
World Health Organization (2022) South Africa Overview. 30 July 2022. https://covid19.who.int/region/afro/country/za
World Health Organization Africa Region (2022) COVID-19 deaths in African region to fall by nearly 94% in 2022: WHO analysis. WHO | Regional Office for Africa June 2, 2022. https://www.afro.who.int/news/covid-19-deaths-african-region-fall-nearly-94-2022-who-analysis
World Population Review (2022) Gini coefficient by country 2022. https://worldpopulationreview.com/country-rankings/gini-coefficient-by-country
WorldOMeter (2022) Coronavirus toll update: cases & deaths by country. Worldometer. https://www.worldometers.info/coronavirus/. Accessed 20 July 2022

5 Re-imagining Informal Settlements as Spaces for Enhancing Women Informal Entrepreneurship in South Africa: A Financial Inclusion Perspective

Purity Hamunakwadi, Sijekula Mbanga, and Frank Moffat

Abstract

Informal settlements are booming in many developing countries through the unprecedented rate at which urbanisation is taking place. Available evidence suggests that the problem lies in the lack of economic opportunities in the formal sector, exacerbating migration to the city, thereby increasing the growth of informal entrepreneurship. Women are found to be at the receiving end of that system since they are excluded and marginalised in the economy. Access to economic opportunities and fostering equitable and sustainable societies is crucial to achieving inclusivity and women empowerment. As such, through an extensive review of literature, this chapter attempts to interrogate and shed light on the role that financial inclusion can play towards supporting women informal entrepreneurship taking lessons from other similar experiences across the world. The work is accomplished on the basis of secondary sources including book reviews, journal articles, research reports, and other secondary materials. Opportunities and barriers to women financial inclusion in African women informal businesses are examined with the aim of providing a possible framework to overcome various impediments mentioned above. The researcher envisions that the findings will be of great benefit to women financial institutions, international agencies, regional bodies, and national and local spheres.

Keywords

Africa · Women · Financial inclusion · Informal entrepreneurship · South Africa

5.1 Introduction

Women informal entrepreneurship has been growing to informal settlement dwellers globally, and most predominantly in Africa. South Africa is not an exception when retrospecting on such issues as most women depend on informal businesses to sustain their livelihoods, yet they struggle because they are financially excluded. According to Menyelim et al. (2021) and Nzotta and Okereke (2009), the financial system significantly enhances economic growth and development. Even so, the problem of limited livelihood

P. Hamunakwadi (✉) · S. Mbanga · F. Moffat
Faculty of Engineering, the Built Environment & Technology, Department of Building and Human Settlements Development, Nelson Mandela University, Port Elizabeth, South Africa
e-mail: Sijekula.mbanga@mandela.ac.za

T. A. Ojo (ed.), *Women and Finance in Africa*, Sustainable Development Goals Series,
https://doi.org/10.1007/978-3-031-53337-2_5

sources is a serious threat to women worldwide, in particular in developing countries, including South Africa (Takaza and Chipo 2022). Moreover, urbanisation has worsened the state since moving to the city is a chance for many men and women to obtain a better life. They aspire for better employment opportunities, larger incomes, safe homes, and access to adequate services in the city. Yet, there are high inequalities and despair that follow city dwellers. In the city, many women as migrants end up dwelling in slums and informal settlements which are associated with poor essential services such as clean water and public transport. Nevertheless, life in the city becomes even more unhealthy and dangerous through various obstacles that impede securing a position in the urban economy (UN Women n.d.). In this regard, women find themselves fighting for their share in the economy.

For women, gender discrimination is a major obstacle and risk factor in fighting for their place in the economy. Hence, linking gender equality and sustainable development is significant because it closes the current gap in wealth between the rich and the poor (Enaifoghe et al. 2021). This inequality gap is problematic to society, so there is a need for attention to address this recurring trend through the realisation of fundamental rights, dignity, and social justice for equity and sustainability of women informal entrepreneurship in the long run. Second, the various effects of socio-economic and environmental pressures that undermine women's role in sustaining their families and communities need to be redressed. Third, and most significantly, building women's capabilities creates better synergies for social and economic transformation through tools such as financial inclusion allocated towards investments in human development priorities (United Nations 2014). This could be an opportunity to invest in according to scholars, since gender equality and sustainable development can effectively be collaborated (Agarwal 1992; Menyelim et al. 2021). Cities and human settlements can be safe and equitable if women who reside in informal settlements are empowered through financial inclusion. In this context, urban planning, governance, and financing sectors must actively implement gender parity measures through continuous participation of women in equal roles of decision-making in an increasingly urban environment (UN Women n.d.). Thus, to ensure gender equality, women must be able to access finance and other financial services without prejudice or exclusion.

According to the United Nations (2018, p. 33;53), "the 2030 Sustainable Development Goals (SDGs) advocate for gender equality and the empowerment of all women and girls (goal 5) as well as making cities and human settlements inclusive, safe, resilient, and sustainable (goal 11)". In this regard, the majority number of women in informal entrepreneurship in Africa reside in informal settlements, where they are excluded from various opportunities. SDG goal 5, target 5a states the importance of "undertaking reforms to give women equal rights to economic resources, as well as access to ownership and control over land and other forms of property, financial services, inheritance, and natural resources, in accordance with national laws" (United Nations 2018, p. 33). This is very crucial to women's well-being, especially in the aspects of access to economic resources and financial services related to this chapter. In addition, goal 11, target 11.7 is as much as applicable to this chapter because it advocates that "by 2030, provide universal access to safe, inclusive and accessible, green and public spaces, in particular for women and children, older persons and persons with disabilities" (United Nations 2018, p. 53). This shows that women need to access green and public spaces safely since their informal businesses are mostly executed in open places. Pertinent to this issue is the double challenge that women face which is residing in informal settlements and being involved in informal entrepreneurship. Thus, a link exists between these two concepts as one mostly influences the other. Against this backdrop, the linkage between informal settlements and women informal entrepreneurship is elucidated and dissected based on global, African, and South African insights and perspectives.

5.2 The Conceptual Link Between Informal Settlements and Women Informal Entrepreneurship

The failure of African urbanisation patterns to resonate with urbanisation theories has long been a source of concern for urban experts in the Global South. The Centre for Housing Rights and Evictions (COHRE) (2008) claims that the urbanisation and the expansion of city ghettos have extraordinary effects, but earlier research in this field has mainly neglected these problems. Weimann and Oni (2019) contend that the nature of cities is reported to have been drastically changed by urbanisation and urban growth throughout the Global South, including Africa. The implication of these trends suggests that the developing world's urbanisation rate is quite rapid, creating what Davis (2004) refers to as "a planet of slums". To back up this assertion, Weimann and Oni (2019, p. 1) indicate that 55% of urban dwellers in sub-Saharan Africa reside in informal settlements, which is greater than the global average of 30%, and other developing countries, including South Asia at 31%. According to the World Bank Group (2016), informal settlements in major cities of South Africa date back to 1940 and 1950 showing that they developed long back. Not only that, but informal settlements in African nations also point to shortfalls in local governance and spatial planning issues which have created an urban form which is characterised by a mix of residential, commercial, and agricultural activities. A mix of residential, commercial, and agricultural activities can be utilised to support localised development which advocates for sustainable human settlements, but they should be well planned, designed, and implemented following spatial policies.

Weimann and Oni (2019) and Chikoto (2009, p. 9) define informal settlements as unauthorised residential areas that do not comply with local authority requirements of building and spatial planning standards with insufficient access to health care and other public amenities. The UN-Habitat (2015, p. 1) defines them based on three main components which are "residential areas without security of tenure, neighbourhoods that lack basic services and city infrastructure and houses not complying with current planning and building regulations mostly situated in geographically and environmentally hazardous areas". Although these two terms are frequently used interchangeably, slums denote to dilapidated housing where residents lack access to portable water and sanitation, security of tenancy, and adequate living conditions. In this chapter, these terms are used interchangeably because residents of those locations have the same difficulties and traits as mentioned earlier. What is appalling in informal settlements is that women frequently work multiple jobs to provide for their families in the informal sector where they occupy informal entrepreneurship activities. Thus, concerning this chapter, it is crucial to define informal entrepreneurship, women entrepreneurship, and women informal entrepreneurship.

Even though many individuals rely on the informal sector, it is frequently ignored. Kinyanjui (2014) argues that a sizable subaltern population that lives in slums relies on the informal economy responding to a robust middle class growing in the cities. Cities Alliance (2021, p. 5) estimates that two billion informal workers and one billion people who live in slums are still at risk of adverse policy environments coupled with social exclusion, poverty, climate change, and deficient public health systems. This implies that although residents of squalid informal settlements experience challenges, they put up with these conditions in the hopes of escaping poverty through employment opportunities (Turok 2015). Hence, informality has become the tool of adaptability utilised by residents in these spaces. It is upon this basis that informal entrepreneurship in developing countries should be regarded as a potential driver of economic development and growth. However, there have been ongoing debates positioning informal entrepreneurship in the wider spaces of the economy that dwindles its operations.

According to Autio and Fu (2014), informal entrepreneurs do not file any incorporation documents or apply for business registration with government authorities, yet they trade legal products and services. On the contrary, women entrepreneurship is any economic activity carried out by women to support themselves through creating new business opportunities, whether formally through registering the company and hiring office space or informally through the use of their knowledge and resources in a home environment (Miah et al. 2017). In this regard, Hundera (2019) cites that to establish and build an entrepreneurial venture, women must first recognise an opportunity, create an idea, and assemble the necessary resources. Thus, women informal entrepreneurship is operating business ventures that are not registered by government authorities to be economically active to earn a living as well as enjoy their economic rights and position in society. The mushrooming of economic informality in developing economies is overwhelming as it is widespread in most African nations.

In the wake of the growing population, the size of the working age is anticipated to increase shortly by 224.0 million by 2050 due to Africa's expanding population and the crucial role played by the urban informal sector (Gluven and Karlen 2020). The informal sector, which is the backbone of economic activity in urban Africa, employs 80.8% of the population. In addition, the informal economy makes a substantial contribution to reducing poverty and is especially prevalent among young people and women, accounting for 95.8% and 92.1% of total employment, respectively (Gluven and Karlen 2020). The debate centres on the claim that women make up a sizable portion of the metropolitan population and that the majority of them work in the informal economy (Ngundu 2010; Chen 2005; Musiolek 2002; Leach 1999). In several countries, the proportion of women working in informal employment exceeds that of men (Women in Informal Employment: Globalising and Organising & International Labour Organisation 2013). Additionally, there is a very strong gender component to the informal economy in slums (Chen 2001). This is because informal economic activities are mainly the only feasible way to earn a living for slum dwellers, hence work as a solution to income generation. Labour informalisation is accompanied by an increase in the number of women in labour force, yet the informal sector activities such as street vending are precarious, unregistered, and lack contracts and social protection (Kinyanjui 2014). In this regard, various other scholars assert that the informal economy is characterised by various economic activities such as hawking/vending, craftsmanship, market trade manufacturing, repairs, hairdressing and beauticians, dressmaking and tailoring, casual employment in unregulated small enterprises, agricultural schemes, and so on (Kinyanjui 2014; Ngundu 2010; Leach 1999). These various activities provide livelihood strategies to generate income for the city dwellers to support their families. Although these livelihood activities generate income for women in slums, they are usually untapped and unrecognised, yet they enable people to survive (Kinyanjui 2014).

Kinyanjui (2014) asserts that while the majority of urban residents rely on the informal economy for living and employment, it is significant to emphasise that the informal economy has been a major source of conflict with modernity and order in the city. In this regard, African cities, in particular, have attempted to combat economic informality without fully comprehending how it operates or the cause of such behaviours. To this end, women live in peri-urban areas and informal settlements where unemployment is rampant. In the same manner, the United Nations Human Settlements Programme (UN-HABITAT) (2006) reveals that African women in urban areas are marginalised due to their gender, physical condition, and socio-economic circumstances. Women's marginalisation, exclusion, and invisibility are exacerbated by planning ideologies and patriarchal structures. According to Kinyanjui (2014), claiming positions in the city and participating in the process of creating the city can be achieved through mobility and solidarity entrepreneurship. Thus, gaining access to space and opportunities for these women is not about opposing male dominance but rather about improving their quality of life and that of their children. In this respect, feminising the city is for identifying livelihood opportunities, sharing spaces, and organising collective action by involving them in all spheres (Kinyanjui 2014).

Collective action may be enhanced by trade associations or other formal unions comprising urban women working in the informal economy (Chant and Mcllwaine 2016). That may assist in furthering women empowerment and transformation at a huge scale. However, the increase of women in the informal sector represents the shortage of opportunities in the formal sector. Most predominantly, gender imbalances are experienced in the region. This is because in African cities, gender inequality is a key concern, especially for the urban poor (Chant and Mcllwaine 2016; Centre on Housing and Evictions 2008). Also, since the process of urbanisation provides very limited options for women to improve themselves, they are captured in the informal sector.

However, they often lack both information and start-up capital (Centre on Housing and Evictions 2008). Also, the national government excludes women's activities in its statistics, yet a variety of jobs including traders, artisians, and suppliers of a range of services can add value to the economic sector. Yet, slum dwellers are involved in various informal economic activities that contribute significantly to the formal economy through the provision of cheap labour and income generation systems. It is argued that informal entrepreneurship absorbs the surplus labour providing income-earning opportunities for the poor (Williams and Youssef 2015). Despite the significant role of women entrepreneurship to informal settlement dwellers, a range of obstacles repeatedly work in contradiction of activities in the informal sector due to limited skills to run substantial businesses. Limited skills in business contribute to the lack of access to critical marketing and capital for start-up investments, business, and employment networks (Women in Informal Employment: Globalising and Organising & International Labour Organisation 2013). According to Hearle et al. (2019), women also have lower endowments in terms of skills, financial services, education, work experience, access to and control over assets, goals, social capital, and networks. In addition, the lack of access to property and finance together with the laws or norms that constrain women from forming independent formal businesses remains a challenge (World Bank 2014; Hearle et al. 2019). According to the International Labour Organisation (2013), women have insufficient knowledge of labour rights, organising, and representation which sideline their bargaining power in the informal economy as compared to men (International Labour Organisation 2013). Thus, social networks are crucial for entrepreneurs because they permit them to expand their businesses and markets and gain access to funding.

According to the International Centre for Research on Women (2019), Hearle et al. (2019), Kabeer (2017), and Miles (2017), women's enterprises are small, less profitable, and less likely to be registered than men. Miles (2017) confirms that these disadvantages mentioned above impede their success because women's credit histories and borrowing power are also impacted by a lack of formal business registration. Also, they limit business growth and legal obligations, rights and protection for employers, and access to business development services. Despite these impediments, financial inclusion can be a pertinent factor in the achievement of women informal entrepreneurship in informal settlements. This is so because women constitute a significant urban population where the majority relies on the informal economy. It is, therefore, imperative to be inclusive in terms of finances towards low-income earners, the poor, and the marginalised. The following section enlightens on the concept of financial inclusion and financial exclusion to give an understanding of what it entails.

5.3 Conceptualising Financial Inclusion

Financial inclusion has grown so much interest and attention worldwide to the extent that the Global Partnership for Financial Inclusion (GPFI) reflects on its potential to contribute towards economic and business development (Global Partnership for Financial Inclusion 2017; Nanziri 2016; Yoshino and Morgan 2016). Financial inclusion is argued to promote inclusive growth and improved income equality, especially for the underprivileged in society. Access to financing has been recognised as one of the fundamental principles of the global development plan at the G20 summit in Seoul. There, the GPFI and a detailed Financial Inclusion Action Plan (FIAP) were developed. According to the Global Partnership for Financial Inclusion (2017), globally, the GPFI has a mandate to advance financial inclusion through increasing usage of and quality access to sustainable formal financial services (GPFI 2017). In the same vein, the World Bank emphasises the need to promote financial inclusion, expansion of financial access, and provision of new data to bolster arguments that financial access improves economic performance (Demirguc-Kunt et al. 2008). In light of the above, defining financial inclusion from various lenses is significant to establish the conceptual understanding that forms the basis of the discussion ahead.

According to Yoshino and Morgan (2016, p. 3), financial inclusion is the extent to which impoverished households and businesses, particularly small- and medium-sized enterprises (SMMEs), have access to financial services. In an extension of the above definition, financial inclusion is a process that encourages accessible, functional, and cheap access to various regulated goods and services for all people and businesses around the world (Mndolwa 2017; Atkinson and Messy 2013). In light of the aforementioned, financial inclusion expands the use of these goods and services by all societal segments through the application of both established and novel strategies, such as financial awareness and education to support financial, economic, and social well-being (Yoshino & Morgan 2016). In the same vein, Hannig and Jansen (2010) assert that financial inclusion aims to bring the "unbanked" into the established financial system so they may readily access a variety of financial services, such as accounts, loans, retirement benefits, insurance, stock markets, payments, and savings. Thus, this chapter adopts the definition that emphasises the degree to which impoverished households and firms, SMMEs access financial services. In this regard, SMMEs represent women informal businesses. According to Mothobi et al. (2020), efforts to increase the number of banking services available to low-income earners and the reduction of dormant accounts can contribute to financial inclusion by increasing access, affordability, consumer financial education, diversification, simplicity, and innovation. Yoshino and Morgan (2016) set a tone that the percentage of adults with personal or dual accounts at recognised financial institutions, such as central banks, credit unions, private institutions, cooperatives, or microfinance institutions (MFIs) (financial regulatory agencies, non-governmental organisations (NGOs), or mobile payment services, has been identified as key indicators of household access to finance. This will help in widening the financial services provided so that low-income earners and the marginalised (women) may have various options to acquire these services. This insinuates that financial inclusion is about providing vulnerable people with financial services that are adequate, appropriate, and affordable through savings, loans, and remittances. Since the emphasis is on recognised financial institutions, suppliers of informal financial services are not included in this context. Also, it is paramount to note that financial inclusion does not disregard the risks and other costs that emerge from services, which on its own may exclude a household or small firm from using their services despite unrestricted access (Yoshino and Morgan 2016). In this regard, these services should at least have reasonable costs that are affordable and encourage usage by low-income earners so that such restrictions will not exclude them from benefiting from financial services. To this end, the following paragraph clarifies the aspect of financial exclusion.

Yoshino and Morgan (2016) explain that financial exclusion (unbanked), on the other hand, is the practice of not using recognised financial institutions, such as banks and microfinance institutions (MFIs) for financial services. Chitimira and Ncube (2020) contend that financial exclusion is the inability of individuals to access and use banking services at reasonable rates in any particular country. Thus, according to Chitimira and Ncube (2020), financial exclusion refers to the prohibition, disqualification, and exclusion of specific individuals from accessing banking facilities including loans and insurance.

The need to differentiate between the two arises from the fact that some people do not require or want to use such services while others might wish, but high costs, lack of resources, lack of trust, ignorance and lack of understanding, or identity requirements hamper their chances to use such services effectively. According to Chitimira and Ncube (2020), it is referred to as voluntary and involuntary financial exclusion, respectively. However, financial exclusion is a concern for both rich and emerging economies, but it is more severe in poor countries (Mothobi et al. 2020; Dymski 2005). Therefore, it is crucial to emphasise that even in some parts of the world with highly developed financial systems, gaps that hamper the success of financial inclusion are still prevalent, especially for low-income earners (Lenka and Barik 2018). Given these flows, robust measures and policies by the state, financial institutions, and other relevant stakeholders are very significant to curb financial exclusion and consider including the low-income earners, in this context women who are into informal entrepreneurship. This will help in promoting equity, social justice, and sustainable transformation of disadvantaged women in the society. In this respect, social justice and equity need to be prioritised for women in informal entrepreneurship to lessen the gender inequality battle that can work against realising the SDG 5 of empowering women and girls. To draw lessons for South Africa, the next section provides more information on gender disparities in terms of access to funding in Africa.

5.3.1 Barriers of Women Informal Entrepreneurship and Financial Inclusion: A Gender Gap Analysis

Gender prejudice works as a barrier towards global progress (Bhatia and Singh 2019). There is so much emphasis on the significance of involving the financial industry in ensuring gender economic justice. Bhatia and Singh (2019, p. 183) assert that “since 2011, 1.2 billion adults worldwide have had access to banking services, but 1.7 billion remain without a bank account”. The low-income households and the unemployed represent half of the population without a bank account. Hence, the G20 summit concluded that financial accessibility is a critical pillar of global progress in 2010 after giving it significant attention. Furthermore, Bhatia and Singh (2019) cite that more than 60 nations have started national initiatives since 2010 with the goal of obtaining universal financial access by 2020. Yet, post-2020, various countries are still battling with issues of financial inclusion to support their populaces. In this respect, funding has been listed as an enabler of 7 goals out of the 17 SDGs. In this regard, providing affordable financial services to women in informal settlements who are involved in informal entrepreneurship is still not recognised and prioritised, yet such mechanisms can be used as strategies towards women empowerment and gender equality for global progress towards achieving SDGs by 2030.

Ojo (2020) confirms that globally the battle to close the gap of financial inaccessibility between men and women has been ongoing. In addition, despite financial inclusion’s role to provide meaningful and inexpensive access to financial services to all individuals and enterprises, implementation is key. Mndolwa (2017) affirms that when it comes to using and having access to financial services, women continue to be discriminated against in sub-Saharan African countries, Pakistan, Bangladesh, and Turkey. Despite substantial advancements in improving access to and usage of recognised financial services by various parts of society, it has been established that women still lag behind in terms of financial support. Until now, there are concerns because women’s significant contribution to economic progress has been erased by their disproportionate exclusion from accessing funding. To this end, women have a lot of potential to advance and close gender disparities through accessing mobile money services. According to the UN Women, encouraging digital and financial inclusion is a key factor in enhancing gender equality and achieving SDG5.

The importance of closing the gender financial disparities has been highlighted in recent literature, leading to the international community to call for increased access for women to financial services. In light of this, policy implementation needs to be considered to achieve gender-equitable financial inclusion on a global scale as several studies demonstrate gender gaps at regional, national, and local variations (Alliance for Inclusion 2016). Experiences of various countries experiencing financial exclusion are noteworthy in this regard since there are notable gaps identified. It is upon these experiences from various countries in which South African women involved in informal entrepreneurship at the same time residing in informal settlements be suggested to be included in financial inclusion mechanisms.

In Rwanda, the informal sector is financed by relatives and friends. Mothobi et al. (2020) cautions that the informal sector's access to formal financial services is hampered by many barriers that hamper the poor, low-income earners, and unregistered business owners from using official banking services. They include but are not limited to lack of credit history, lack of knowledge and expertise required to prepare financial statements, insufficient collateral, and weak business models and plans. Women in Tanzania share similar constructs with other African nations in which they are more reliant, less educated, poorer, and have lower employment rates than men. In this regard, because more women than males lack formal education, the possibility of women accessing formal financial accounts has reduced by 58.4% (Mndolwa 2017, p. 27). Mndolwa (2017, p. 27) estimates that employment increases the rate at which women are financially included by 25%, yet fewer women are formally employed. Also, women historically and today have lower levels of financial and digital knowledge and are less likely to own mobile phones that can be used to access mobile money accounts. This demonstrates unequivocally that the main obstacles preventing women from fully achieving financial inclusion are still unemployment, financial, and digital illiteracy.

Studies at the country level in African and Asian regions have examined the gender differences in financial inclusion. Aterido et al. (2013) affirm that in nine sub-Saharan African (SSA) nations, including South Africa, Botswana, Namibia, Rwanda, Kenya, Malawi, Tanzania, Zambia, and Uganda, examination of the gender divide in the usage of formal banking services and informal financial services by individuals was conducted. Similarly, Mndolwa (2017) confirms that according to research, women are typically 14% less likely than males to own a mobile phone in low- and middle-income economies. In low- and middle-income nations, more than 1.7 billion women have no ownership of mobile phones, and even among those who do, there is a large gender difference in mobile phone usage. Five major obstacles to mobile use were cited by women in low- and middle-income nations. The most significant of them was the price of using mobile phones, while the others were harassment, poor network coverage, agent pushback, and lack of technological skills to use digital services. Therefore, there is a greater need demand for policies that can help mobile money users achieve affordability and financial literacy.

Since the adoption of the financial sector reforms under the Financial Sector Charter of 2003, the South African government and financial organisations have not yet considered access to funding as an effective strategy for improving the status of women (Ojo 2020; Nanziri 2016). Ojo (2020) further states that women's financial illiteracy, which results in them lacking financial knowledge and abilities with regard to budgeting, investing, paying taxes, and borrowing, has an impact on how banking services (savings and investments) are understood by South Africans. Chitimira and Ncube (2020) affirm that financial consumers in particular the poor and low-income earners who dwell in informal settlements and remote regions and are informally employed are not financially educated. Thus, financial illiteracy refers to the ability to make informed decisions by individuals with regard to managing their finances either on short-term or long-term basis (Chitimira and Ncube 2020). Hence, financial consumers become financially illiterate when they lack the fundamental information, education, skills, and understanding of the financial services that are accessible to them, preventing them from making wise choices regarding credit borrowing, personal finance, and future investments. This implies that in South Africa, many people are still financially illiterate, which is an impediment for them to enjoy financial products and services offered by formal financial institutions. In this respect, financial illiteracy is working against creating a sustainable environment for female entrepreneurs since the knowledge that is lacking will help them to make informed decisions.

In South Africa, financial exclusion of the needy is exacerbated by excessive documents required in the process of opening bank accounts (Chitimira and Ncube 2020). These documents are strictly required for compliance purposes through "know your customer", KYC, which include proof of income, bank statements with financial records that reflect the amount before opening new accounts, identity documents, and proof of residence, which may be very difficult to be owned by the poor and low-income earners (Chitimira and Ncube 2020). If this is the case for those who earn low incomes, how much more will the same issues impact people living in informal settlements without recognised or official addresses?

Apart from the fact that the needy in South Africa are excluded from the financial system, a contrasting narrative suggests that they could have bank accounts and still be financially excluded. In this way, South Africa differs from other African nations in that women possess a greater proportion of banks than men (Ojo 2020). However, social grants through the South African Social Security Agency (SASSA) are the main cause of the larger percentage of access to banking services by women in South Africa. Ojo (2020, p. 42) and Fanta and Mutsonziwa (2016) assert that out of the entire population surveyed, 64% represents individuals with bank accounts while 74% represents those with SASSA Mastercard, including the number of those with their names on bank accounts (Fanta and Mutsonziwa 2016). According to the study, 13% have SASSA cards alone (Ojo 2020).

While women in South Africa have access to financial services (bank ownership), just 28% of them make use of their bank accounts, while a large majority have inactive accounts (Ojo 2020, p. 43). Bank accounts were divided into four categories by the United Nations Capital Development Fund (UNCDF): used accounts, with at least three transactions per month, inactive accounts, with about two transactions per month, and active accounts, with two transactions per month, and inactive accounts, with no transactions per month. However, while most South African women own banks and are therefore financially engaged, they are excluded financially since they do not use banking facilities to save and invest. Also, they lack the financial records required to obtain funding. It can be argued that owning a bank account does not guarantee access to finance from financial bodies. Hence, it also suggests that financial exclusion can be in two forms which are non-bank owners and bank owners but not financially active.

Mothobi et al. (2020, p. 10) posit that many people (70%) in South Africa work in the informal sector, whereas 61% of Rwandans and 60% of Tanzanians in the same sector were not educated; however, 56% of Mozambicans completed their primary school. Interestingly, in these three countries mentioned above, men still dominate ownership of the informal sector. These stereotypes result from various problems such as religious and traditions that discourage women from entrepreneurship. Specifically, in South Africa, despite the fact that a sizable majority of workers from informal enterprises are women, businesses are still predominantly owned by men. Hence, they reported that they do not keep financial records, nor do they separate business income from personal finances (Mothobi et al. 2020).

Although South Africa has a sophisticated financial system that offers a variety of loan options, just 2% of unregistered enterprises can obtain authorised financing, leaving informal entrepreneurs in a difficult situation (Mothobi et al. 2020:10). Additionally, despite South Africa having one of the most developed markets out of all the nations studied, financial inclusion for informal businesses is quite poor (Mothobi et al. 2020). According to a FinFind research that examined 11,033 small business funding requests in 2017, only 1% of SMMEs in South Africa have received funding, and 87% were self-funded

(FinFind 2018, p. 13). Similarly, in Senegal only 2% of informal business owners have access to formal bank loans where Ghana, Kenya, Rwanda and Tanzania were all at 1% (Mothobi et al. 2020, p. 12). Also, Mothobi et al. (2020, p. 12) affirm that results from estimates from Registered Investment Advisor (RIA) survey data suggested that 2% of South African small business owners are financially involved and 89% are either self-funded or unbanked. In the nine countries covered by RIA survey, only 6% of informal businesses use business accounts, while 18% utilise owners' bank account (Mothobi et al. 2020, p. 20). In this survey, only 3% of the participants claimed to have obtained a bank loan and only 4% reported having a credit card (Mothobi et al. 2020, p. 20). The lack of collected data and regulations governing credit record data makes it difficult for SMMEs to acquire loans from banks. Likewise, businesses in the informal sector have a high possibility of not having these credit records, despite recognised consumer data in South Africa. Unfortunately, all these incidents make it more difficult for unregistered businesses to benefit from the regulated financial system.

Women in South Africa continue to experience extreme poverty, insecurity, and inequality regardless of established funding organisations and policies that advocate for the emancipation of women. This suggests that even though financial inclusion can contribute to economic and financial development, there are still some significant gaps that have slowed the empowerment process. Evidence is shown by low-income earners and the underprivileged in the society not accessing such services. The next section provides an exposition on the significance of financial inclusion to women while drawing lessons from other comparable examples across the globe.

5.4 Why Women's Financial Inclusion Is Important: Lessons for South Africa

Financial inclusion is of much significance to women informal entrepreneurs who reside in informal settlements and slums because they have double informality challenges. Kyungha (2022) states that prioritising policies that help women to access funding reduces the rate at which they are financially excluded. Access to finance can lead to positive economic outcomes as evidenced by a body of empirical studies (Mndolwa 2017). Thus, the influence of financial inclusion on women empowerment is suggested to be an instrument that can help to achieve financial development and reduce income inequality.

Examples are shown by studies that focussed on the links between South Asian slums and the empowerment of women (Kaur et al. 2015; Nasrin 2012; Hazarika 2010; Fisher 2008). Thus, the same approach can be used in reference to South African informal settlements on the verge of acquiring women empowerment and gender equality. A reference study of 737 females was conducted on the influence of financial inclusion on the empowerment of women who were slum dwellers in India (Hazarika 2010). The results show that schemes such as Pradhan Mantri Jan Dhan Yojana (PMJDY), Pradhan Mantri Jivana Jyoti Bima Yojana (PMJJBY), Pradhan Mantri Suraska Bima Yojana (PMSBY), and Atal Pension Yojana (APY) contributed to financial increase and empowerment of women especially to those who possessed and used their bank accounts (Hazarika 2010). The findings also showed that the PMJDY initiative had a beneficial effect on the socio-economic and political spheres for women living in slums (Hazarika 2010). This suggests that owning a bank account and using the bank account helped women in India to be financially included, and as a result, their socio-economic and political dimensions were positively influenced.

In the 1990s, a study of three major financial initiatives that was conducted in Bangladesh for women revealed that female customers improved their household's livelihoods after using banking services, and 5% of female customers escaped from poverty each year by participating in microfinance programmes (Kyungha 2022, p. 309). Swamy (2014, p. 12) examined the effects of women's engagement in self-help groups (SHGs) with microfinance between 2007 and 2012 in India. Women benefited more from SHGs than men with a significant increase on income shown by 8.4% and 3.9%, respectively. In addition, there was an increase in investment of social and human capital resulting from the rise of income. Likewise, according to Beaman et al. (2014, p. 3), a study on the effects of financial inclusion in Mali which involved a household panel survey of 6000 families confirm that access to savings clubs increased savings by 30% and credit availability by 3%. Thus, if women are able to save up to 30%, it can help them to support their livelihoods.

Prina (2013, p. 16) performed a field experiment in 19 slums of Nepal for random sampling of 1118 female-headed households who were given bank accounts without opening, maintenance, or withdrawal fees. This approach was reported to be ideal as 80% out of 84% of women who created accounts saved while depositing at least twice in the first year. This demonstrates how having a bank account and using it regularly supports the practice of saving. In addition, Kyungha (2022, p. 309) confirms that another study in Nepal on emancipating women discovered that 68% of women who accessed financial opportunities contributed to decision-making processes in crucial developments of society such as buying and selling of properties, preparing family events (children's weddings and educational support). Similarly, Dobra (2011) affirms that in Tanzania, regardless of the size of the business, the Women Entrepreneurship Development Trust Fund (WEDTF) gained more respect and emancipation through starting a business. According to a survey conducted in Rwanda, 38% of women who used Urwego Opportunity Bank (UOB) developed more advanced ideas of business management, and 69% of female clients gained self-respect (Kyungha 2022, p. 309). Presumably, their financial engagement capacitated women to gain more social authority and confidence which enabled them to assume more responsibility for household and societal decisions and community initiatives. Hence, to support the above claims, Kabeer (2009) asserts that in comparison to men, women use resources more efficiently so they can better support their families.

Financial inclusion of women in Ghana and Bolivia also improved their status and increased their self-confidence in society. Women's economic empowerment contributed to independence, self-determination, and support to their children. Evidence show that children of microfinance clients attended school, stayed in school for longer periods, and experienced less child labour with low rates of student dropouts (Kyungha 2022). Also, a 1996 survey of children in Grameen Bank villages had better chances to attain education as compared to other children in villages who were non-members. In this regard, high percentages were observed, such as 81% of males in Grameen homes who were enrolled for education as opposed to 54% of boys in non-Grameen households, and all girls in Grameen homes were enrolled for education. All of these instances demonstrate the potential of financial inclusion as an intervention against a range of social ills.

Women who are financially active have steady spending patterns and improved incomes which improve food security, health, and living conditions. According to a USAID-AIMS study in Uganda, it was discovered that 32% of customers participated in at least one AIDS prevention practice, compared to 18% of non-customers. Also, 95% of female customers in the Foundation for Credit and Community Assistance (FOCCAS) have improved the health and food security of their children (Kyungha 2022, p. 310). This demonstrates how these programs can increase women's participation and engagement to their advantage. Since women's utilisation and access to financial services encourage both socio-economic and personal development, the twin concerns of informality informal settlements and women informal entrepreneurship can be addressed using the same strategy. In this regard, bridging the gap in gender inequality in Africa for achieving SDGs is significant to enjoy social justice and opportunities across all facets of society (Enaifoghe et al. 2021). Breaking forth that limitation of gender inequality, there is a need to increase economic opportunities for women and re-imagine some customs and practices that are entrenched in primitive laws and policies to tally with international laws and treatises such as the SDGs 2030. Thus, women financial inclusion needs to be engraved in various municipality plans and local government levels to ensure that

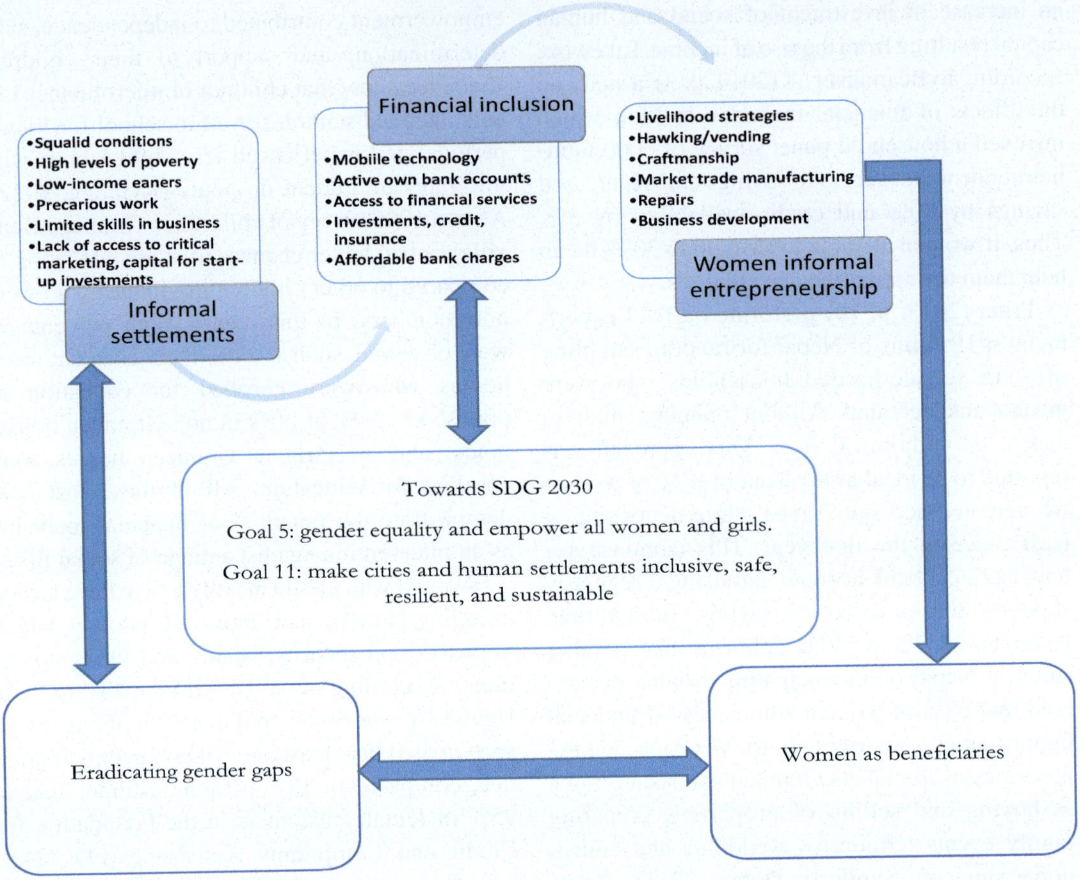

Fig. 5.1 Proposed framework for women informal entrepreneurship and financial inclusion. (Source (s): Authors' compilation)

inclusive development is prioritised. The following proposed framework shows the linkages between informal settlements, financial inclusion, and women informal entrepreneurship towards bridging the gaps of gender inequality to achieve SDGs 2030 (Fig. 5.1).

5.5 Conclusion

The chapter examined ways to empower women in informal settlements through financial inclusion in their small informal businesses, relying on examples from around the globe of similar initiatives. Numerous observations showed that financial inclusion enables small and new enterprises to access a variety of financial products including risk management, payment, and credit services as well as savings and investment opportunities. Since informal businesses are unregulated, have inactive accounts, lack credit history, and lack financial literacy, their demise is more appalling. Despite having one of the most extensive financial systems, developed markets, and the most extensive regulated data, South Africa, nevertheless when referring to informal businesses, still faces similar challenges in the region. South Africa differs in that the majority of women are financially included by having bank accounts, but in practice, they are excluded from gaining rewards due to owning of dormant accounts. Hence, evidence from the literature demonstrated that it was resulting from owning businesses in the unregulated and unregistered informal sector. Since banks and other formal financial institutions rely on credit record, it is recommended that information technology serve as a catalyst for financial inclusion. Financial technologies (fintech) which improve financial inclusion and lower information inconsistencies and transaction costs can break down that barrier. Therefore, mobile banking services and other digital services may produce information which is worth considering in evaluating credit viability as unregistered enterprises frequently lack insurance and credit histories.

References

Agarwal B (1992) The gender and environment debate: lessons from India. Fem Stud 18(1):119–158. https://doi.org/10.2307/3178217

Alliance for Financial Inclusion (2016) Denarau action plan: the AFI network commitment to gender and women's financial inclusion. www.afi-global.org/sites/default/files/publications/2016-09/Denarau%20Action%20Plan.pdf

Aterido R, Beck T, Iacovane L (2013) Access to finance in sub-Saharan Africa: is there a gender gap? World Dev 47(C):102–120. https://doi.org/10.1016/j.worlddev.2013.02.013

Atkinson A, Messy F (2013) Promoting financial inclusion through financial education: OECD/INFE evidence, policies and practice, OECD working papers on finance, insurance and private pensions, no. 34, OECD publishing, Paris. https://doi.org/10.1787/5k3xz6m88smp-en

Autio E, Fu K (2014) Economic and political institutions and entry into formal and informal entrepreneurship. Asia Pac J Manag 32(1):67–94. https://doi.org/10.1007/s10490-014-9381-0

Beaman L, Karlan D, Thuysbaert B (2014) Savings for a (not so) rainy day: a randomised evaluation of savings groups in Mali. (economic growth center discussion paper no.1043) economic growth center (ECG), Yale University. https://ssrn.com/abstract=2508450

Bhatia S, Singh S (2019) Empowering women through financial inclusion: a study of urban slum. VIKALPA J Decis Makers 44(4):182–197. https://doi.org/10.1177/0256090919897809

Centre on Housing Rights and Evictions (2008) Women, slums and urbanisation: examining the causes and consequences. Geneva. COHRE. Women and Housing Rights Programme. https://www.globalinitiative-escr.org/wp content/uploads/2013/05/women_slums_and_urbanisation_may_2008.pdf

Chant S, Mcllwaine C (2016) Cities, slums and gender in the global south: towards a ferminised urban future. Routledge, Abingdon

Chen M (2001) Women in the informal sector: A global picture, the Global movement. SAIS Rev 21:71–82. https://doi.org/10.1353/sais.2001.0007

Chen MA (2005) Rethinking the informal economy: linkages with the formal economy and the formal regulatory environment, WIDER Research Paper, No. 2005/10, ISBN 9291906891, The United Nations University World Institute for Development Economics Research (UNU-WIDER), Helsinki. https://www.econstor.eu/bitstream/10419/63329/1/488006546.pdf

Chikoto T (2009) Informal settlements in South Africa. [Honours dissertation, University of Pretoria]. Pretoria https://repository.up.ac.za/bitstream/handle/2263/14436/Chikoto_Informal(2009).pdf?sequence=1

Chitimira H, Ncube M (2020) Legislative and other selected challenges affecting financial inclusion for the poor and low-income earners in South Africa. J Afr Law 64(3):337–355. https://doi.org/10.1017/S0021855320000182

Cities Alliance (2021) The challenge of slums-an overview of past approaches to tackle it. Cities Alliance, Brussels. https://www.alexandria.unisg.ch/server/api/core/bitstreams/af014d57-7124-4c09-9137-f4d0261c09aa/content

Davis M (2004) Planet of slums: urban involution and the informal proletariat. New Left Rev 26:5–34. https://jbburnett.com/resources/davismike_nlr26001.pdf

Demirguc-Kunt A, Beck T, Honohan P (2008) Finance for all? Policies and pitfalls in expanding access. The International Bank for Reconstruction and Development. World Bank, Washington, DC. https://www.researchgate.net/publication/254436993_Finance_for_All_Policies_and_Pitfalls_in_Expanding_Access

Dobra A (2011) Microfinance: champion in poverty alleviation and failure in female empowerment. Internationale Politic and Gesellscaft 3:134–144. https://library.fes.de/pdf-files/ipg/2011-3/12_dobra.pdf

Dymski GA (2005) Banking strategy and financial exclusion: tracing the pathways of globalisation. Rev Econ 31(1–29):107–143. https://doi.org/10.5380/RE.V31I1.5021

Enaifoghe AO, Brown C, Maramura TC, Ajede SA (2021) Bridging the inequality gap in Africa for sustainable development: assessing the role of women. Int J Dev Sustain 10(7):204–279. https://www.researchgate.net/publication/359636256_Bridging_the_inequality_gap_in_Africa_for_sustainable_development_Assessing_the_role_of_women_GENDER_BASED_STUDIES_View_project_South_Africa%27s_Management_and_Investments_in_SADC_A_Strategy_for_the_Imp

Fanta AB, Mutsonziwa K (2016) Gender and financial inclusion analysts of financial inclusion of women in the SADC region. Policy research paper no.01/2016. FinMark trust. South Africa. https://doi.org/10.13140/RG.2.1.1390.3605

FinFind (2018) Inaugural South African SMME access to finance report. https://www.busa.org.za/wp-content/uploads/2018/11/FinFind-SA-SME-Inaugural-SMME-Access-to-Finance-Report-2018-1.pdf

Fisher J (2008) Women in water supply, sanitation and hygiene programmes. Proc ICE Munic Eng 16(4):223–229. https://doi.org/10.1680/muen.2008.161.4.223

Global Partnership for Financial Inclusion (2017) Financial inclusion action plan, July 2017. https://www.gpfi.org/sites/gpfi/files/documents/2017%20G20%20Financial%20Inclusion%20Action%20Plan%20final.pdf

Gluven M, Karlen R (2020, December 3) Supporting Africa's urban informal sector: coordinated policies with social protection at the core. World Bank Blogs https://blogs.worldbank.org/africacan/supporting-africas-urban-informal-sector-coordinated-policies-social-protection-core

Hannig A, Jansen S (2010) Financial inclusion and financial stability: current policy issues, ADBI working paper 259. Asian Development Bank Institute, Tokyo. http://www.adbi.org/working-paper/2010/12/21/4272.financial.inclusion.stability.policy.issues/

Hazarika I (2010) Women's reproductive health in slum populations in India: evidence from NFHS-3. J Urban Health 87(2):264–277. https://doi.org/10.1007/s11524-009-9421-0

Hearle C, Baden S, Kalsi K (2019) Promoting economic empowerment for women in the informal economy. Work and Opportunities for Women (WOW) Helpdesk Guidance No1 Final Draft, September 2019. https://assets.publishing.service.gov.uk/government/uploads/system/uploads/attachment_data/file/844320/Promoting-Economic-Empowerment-Women-Informal-Economy-Oct19.pdf

Hundera MB (2019) Role conflict, coping strategies and female entrepreneurial success in sub-Saharan Africa. [PhD Thesis, Tilburg University]. CentER, Centre for Economic Research. https://doi.org/10.26116/center-lis-1921

International Centre for Research on Women (2019) The benefits of collective bargaining for women: a case study of Morocco, Solidarity Centre. https://www.solidaritycenter.org/wp-content/uploads/2019/02/The-Benefits-of-Collective-Bargaining-for-Women-A-Case-Study-of-Morocco.219.pdf

International Labour Organisation (2013) The informal economy and decent work: a policy resource guide-supporting transitions to formality. International Labour Organisation https://www.ilo.org/wcmsp5/groups/public/%2D%2D-ed_emp/%2D%2D-emp_policy/documents/publication/wcms_212689.pdf

Kabeer N (2009) World survey on the role of women in development: women's control over economic resources and access to financial resources, including microfinance, economic and social Commission for Asia and the Pacific (ESCAP) 12–14 November 2008, United Nations division for the advancement of Women Department of Economic and Social Affairs, New York. https://www.un.org/womenwatch/daw/ws2009/documents/EC%20Report-%2026FEB09.pdf

Kabeer N (2017) Women's economic empowerment and inclusive growth. Labour markets and enterprise development. GROW working paper series GWP-2017-01-concept paper, Institute for the Study of international development (ISID). http://grow.research.mcgill.ca/publications/working-papers/gwp-2017-01.pdf

Kaur M, Singh S, Gupta M, Bahuguna P, Rani S (2015) Inequity in access to health services between migrants and natives of Chandigarh, India. Int J Migr Health Soc Care 11(2):147–155. https://doi.org/10.1108/IJMHSC-03-2014-0010

Kinyanjui MN (2014) Women and the informal economy in urban Africa: from the margins to the Centre. Zed Books, London

Kyungha K (2022) Assessing the impact of mobile money on improving the financial inclusion of Nairobi women. J Gend Stud 31(3):306–322. https://doi.org/10.1080/09589236.2021.1884536

Leach F (1999) Women in the informal sector: the contribution of education and training. In: Rowan-Campbell D (ed) Development with women. Oxfam Publication, Oxford, pp 46–62

Lenka SK, Barik R (2018) A discourse of financial inclusion: post-liberalisation mapping in rural and urban India. J Financ Econ Policy 10(3):406–425. https://doi.org/10.1108/JFEP-11-2015-0065

Menyelim CM, Babajide AA, Omankhanlen AE, Ehikioya BI (2021) Financial inclusion, income inequality and sustainable economic growth in sub-Saharan African countries. Sustain For 13(1780):1–15. https://doi.org/10.3390/ma14040753

Miah J, Khan SA, Misto OF, Karim R (2017) Women entrepreneurship of Sylhet City, Bangladesh: challenges and opportunities. Manag Organ Stud 5(3):17–28. https://doi.org/10.5430/mos.v5n3p17

Miles KS (2017) Case study: gender-sensitive BER and informality, Business Environment Working Group, Donor Committee for Enterprise Development. https://www.enterprise-development.org/wp-content/uploads/DCED-BER-Gender-Case-Study-Informality.pdf

Mndolwa FD (2017) Determinants of gender disparities in financial inclusion: insights from Tanzania. [Master's dissertation, University of Cape Town]. Cape Town. https://open.uct.ac.za/bitstream/handle/11427/27332/thesis_com_2017_mndolwa_florence_d_SR.pdf?sequence=1&isAllowed=y

Mothobi O, Gillward A, Aguera P (2020) A demand side view of informality and financial inclusion (policy paper no.10, series 5: after access-assessing digital inequality in Africa). Research ICT Africa https://www.researchictafrica.net/wp/wp-content/uploads/2020/02/RIA-Small-and-Micro-business-Report-2019.pdf

Musiolek B (2002) Decent work in the informal sector: CEE/CIS region. Working paper on the informal economy. Geneva, International Labour Organisation. https://www.ilo.org/wcmsp5/groups/public/%2D%2D-ed_emp/documents/publication/wcms_122063.pdf

Nanziri EL (2016) Financial inclusion and welfare in South Africa: Is there a gender gap? J Afr Dev 18(2):109–134. https://doi.org/10.5325/jafrideve.18.2.0109

Nasrin F (2012) Women, environment and environmental advocacy: challenges for Bangladesh. Asian J Soc Sci Humanit 1(3):149–172. http://www.ajssh.leena-luna.co.jp/AJSSHPDFs/Vol.1(3)/AJSSH2012(1.3-16).pdf

Ngundu K (2010) The impact of the informal economy on the social and economic development of women headed households in Chegutu urban district in Zimbabwe. [Master's dissertation, University of Pretoria]. Pretoria. https://repository.up.ac.za/.../27370/dissertation.pdf

Nzotta SM, Okereke EJ (2009) Financial deepening and economic development of Nigeria: an empirical investigation. Afr J Account Econ Finance Banking Res 5(5):52–66. https://www.researchgate.net/publication/228303494_Financial_Deepening_and_Economic_Development_of_Nigeria_An_Empirical_Investigation

Ojo TA (2020) A study of financial inclusion and women's empowerment in South Africa: the case of female entrepreneurs in Gauteng. [PhD Thesis, University of Pretoria]. Pretoria. https://repository.up.ac.za/.../27370/dissertation.pdf

Prina S (2013) Banking the poor via saving accounts: evidence from a field experiment. J Dev Econ 115:16–31. https://doi.org/10.1016/j.jdeveco.2015.01.004

Swamy V (2014) Financial inclusion, gender dimension, and economic impact on poor households. World Dev 56:1–15. https://doi.org/10.1016/j.worlddev.2013.10.019

Takaza SC, Chipo C (2022) An investigation into the livelihoods strategies of informal women traders in Zimbabwe. J Innov Entrep 11(37):1–22. https://doi.org/10.1186/s13731-021-00175-3

Turok I (2015) Upgrade informal settlements: South Africa. S Afr J Soc Econ Policy 57:10.10520/EJC168940

UN Women (n.d.) Women and sustainable development goals, United Nations entity for gender equality and the empowerment of women, 22–23. https://africa.unwomen.org/sites/default/files/Field%20Office%20Africa/Attachments/Publications/2016/03/Analysis%20of%20Women%20and%20Sustainable%20Development%20Goals-%20Web%20Rev.pdf

UN-HABITAT (2006) Nairobi urban sector profile. United Nations Human Settlements Programme. UN-HABITAT, Nairobi. https://www.zhiyanbao.cn/index/partFile/5/unhabitat/2022-03/5_19339.pdf

UN-HABITAT (2015) Habitat III issue paper 22- informal settlements. New York. https://unhabitat.org/sites/default/files/download-manager-files/Habitat-III-Issue-Paper-22_Informal-Settlements-2.0%20%282%

United Nations (2014) World survey on the role of women in development 2014. Gender equality and sustainable development. UN Women, (a/69/156), New York. https://www.unwomen.org/sites/default/files/Headquarters/Attachments/Sections/Library/Publications/2014/World-survey-on-the-role-of-women-in-development-2014-en.pdf

United Nations (2018) The 2030 agenda and the sustainable development goals: an opportunity for Latin America and the Caribbean. (LC/G.2681-P/rev.3), Santiago, 2018. https://www.cepal.org/sites/default/files/publication/files/40156/S1801140_en.pdf

Weimann A, Oni T (2019) A systematised review of the health impact of urban informal settlements and implications for upgrading interventions in South Africa, a rapidly urbanising middle-income country. Int J Environ Res Public Health 16(3608):1–17. https://doi.org/10.3390/ijerph16193608

Williams CC, Youssef YA (2015) Theorising entrepreneurship in the informal sector in urban Brazil: a product of exit or exclusion? J Entrep 24(2):148–168. https://doi.org/10.1177/0971355715586897

Women in Informal Employment: Globalising and Organising., & International Labour Organisation (2013) Women and men in the informal economy. In: A statistical picture, 2nd edn. International Labour Organisation

World Bank (2014) Gender at work: a companion to the world development report on jobs, World Bank. https://documents1.worldbank.org/curated/en/884131468332686103/pdf/892730WP0Box3800report0Feb-02002014.pdf

World Bank Group (2016) Scaling up the upgrading of informal settlements: a scoping study of south African good practices and innovations. Department of National Treasury. https://csp.treasury.gov.za/DocumentsToolbox/CSP%20Upgrading%20Scaling%20Up%20Scoping%20Report.pdf

Yoshino N, Morgan P (2016) Overview of financial inclusion, regulation, and education. ADBI working paper 591. Asian Development Bank Institute, Tokyo. https://www.adb.org/sites/default/files/publication/190672/adbi-wp591.pdf

6 The Role of Institutions in Promoting Financial Inclusion of MSMEs in Zimbabwe

Belinda R. Chaora

Abstract

Institutions play an important role in the delivery of financial products and services and are a vital component of the financial inclusion ecosystem. In Zimbabwe, the definition of financial inclusion is limited to access and use of financial products and services from formal institutions in the financial service sector such as insurance agents, mobile network operators, banks, microfinance institutions, remittance agents and fintech companies. Their oversight, regulation and survival are determined by the policy environment around them. The definition of inclusion, however, does not account for the informal operations which are taking place in Zimbabwe where 52% of micro, small, and medium enterprises (MSMEs) are informal and therefore look to other avenues for access to financial products and services. This chapter processes data from a nationwide survey of MSMEs as primary data and subsequently uses secondary sources to compile other informal institutions influencing the financial habits of MSMEs in Zimbabwe. The chapter outlines some of the barriers experienced by MSMEs when attempting to access products and services from formal sources. It further outlines the key players in financial service provision and their roles noting that other institutions (social and cultural) should additionally be considered in future efforts to understand the state of inclusion and promote financial equity in Zimbabwe.

Keywords

Institutions · Financial inclusion · Financial access · Micro · Small and medium enterprises (MSMEs) · Zimbabwe

6.1 Introduction

The COVID-19 pandemic disrupted social and economic growth on the African continent. It has created (and, in many cases) sharpened already existing vulnerabilities, deepening the levels of inequality in Africa. Although recovery is possible, inequality is a major developmental challenge going forward. In the midst of the myriad of challenges and opportunities for development faced by African leaders, financial inclusion has been determined to be a prime candidate to promote more equitable development by targeting vulnerable bottom of the pyramid (BOP) wealth groups. The World Bank (2022) postulated that post-COVID-19, recovery had started in 2022, and sub-Saharan Africa (SSA) is experiencing a

B. R. Chaora (✉)
SIVIO Institute, Harare, Zimbabwe
e-mail: belindac@sivioinstitute.org

T. A. Ojo (ed.), *Women and Finance in Africa*, Sustainable Development Goals Series,
https://doi.org/10.1007/978-3-031-53337-2_6

renewed increase in rates of gross domestic product (GDP) growth with an average of 4.6% growth compared to the world average of 5.8% as of 2021.

Previous research by the World Bank (2022) has demonstrated that not only is access to financial products and services important for economic growth and financial equity in communities, but the quality of those products and the manner in which they are made available are also important. The Organisation for Economic Co-operation and Development (2018) has stated that the availability of financial products and services is governed by institutions which make regulatory decisions (both good and bad), provide services through direct interface with consumers, develop innovative financial technologies and influence cultural norms to determine how products will eventually be accepted or used.

6.2 Background to Financial Inclusion in Zimbabwe

The Zimbabwean government recognized the impact an inclusive financial service sector will have on improving the socio-economic development of Zimbabwe. The lack of access to bank accounts, insurance and savings facilities was noted in 2012 when only 30% of Zimbabwean adults made use of bank services and only 14% of MSMEs were banked as stated by the Reserve Bank of Zimbabwe (RBZ). This prompted the development of a National Financial Inclusion Strategy – NFIS of 2016 to 2020 (Ministry of Finance, 2020). This strategy saw the engagement of multiple stakeholders in financial inclusion efforts across the financial services sector. The pursuit of equitable growth through financial inclusion is linked with the development agenda for Zimbabwe which was elaborated in the country's Transitional Stabilization Plans (TSP) of 2018 to 2020 (Government of Zimbabwe [GOZ], 2018a), National Development Strategy 1-NDS1 of 2021 to 2025 and Vision 2030 of 2018 to 2030 (GOZ, 2018b). According to the Reserve Bank of Zimbabwe (2016), the NFIS of 2016–2020 revolved around four core pillars, namely microfinance, innovation, consumer protection and financial capability, all of which rely on an institution for their execution. In Zimbabwe, financial inclusion efforts can only be recognized through formal institutions as noted in the definition of inclusion used by the Reserve Bank of Zimbabwe (2016) which defines financial inclusion as "the effective use of a wide range of quality, affordable and accessible financial services provided in a fair and transparent manner through formal /regulated entities by all Zimbabweans."

The Reserve Bank of Zimbabwe further states that the NFIS of 2016 identified the following key sectors in the financial service framework which are instrumental in the delivery of inclusive products, and these are pensions, insurance, capital markets, banking, payment systems, microfinance institutions and development finance institutions as illustrated in Fig. 6.1.

In addition to these pillars, we see that fintech companies and mobile network providers have become significant in promoting inclusion. The presence of players such as Mukuru, City Hopper, Get Bucks and InnBucks has allowed for the movement of funds not just from outside Zimbabwe's borders into the economy through remittances but also within and across the country as money (primarily foreign currency) moves from city to city bypassing the need for formal bank accounts.

The rise of social media is an influence that cannot be ignored given the role that social media channels like WhatsApp, Facebook, Twitter, YouTube and TikTok are now playing in the dissemination of information and influencing trends in the access to financial products and services. According to Statcounter (2022), 35% of Zimbabwean Internet users access Facebook, 24% access Twitter, 17% access YouTube, 15% access Pinterest and 5% access Instagram. This demonstrates that these platforms are growing in their influence on Zimbabwean citizens.

Cultural institutions, which may not be recognized by the Government of Zimbabwe (GoZ) as formal institutions in the financial inclusion discourse, are nonetheless influential. Here the reference is mainly to family and religious institutions whose influence is created in forma-

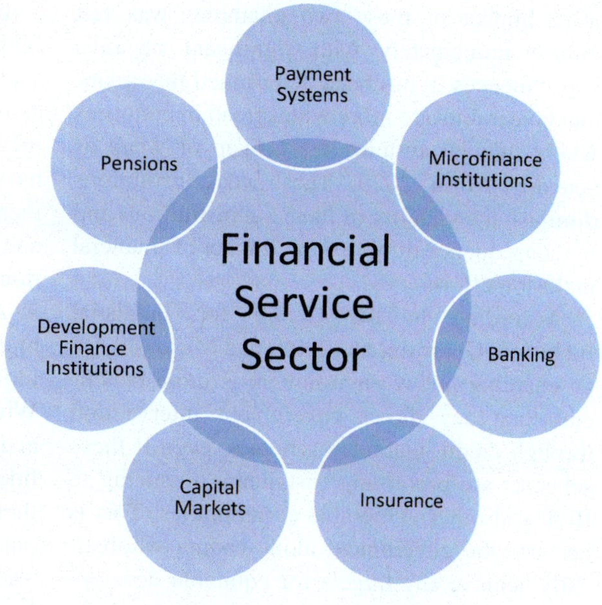

Fig. 6.1 Players in the financial services sector in Zimbabwe. (Source: Reserve Bank of Zimbabwe, National Financial Inclusion Strategy – 2016–2020)

tive life stages and dictates worldviews including those about how money should be accessed, passed on, used and inherited. Murisa (2022) in his research on land tenure in Chilonga has shown that customary tenure still remains in Zimbabwe and is upheld by at least 55% of Zimbabwe's population which is primarily rural.

Over and above the institutions stated above is the overarching institution of the GoZ whose role is to create and implement policies around financial inclusion. GoZ has been outworking this role primarily through the RBZ. The GoZ demonstrates oversight through central government and corresponding line ministries as well as through local authorities whose oversight is divided into geographical regions, i.e., cities, towns or districts.

6.2.1 Institutions and Policy

It is the responsibility of governments to create policies which form preconditions for the financial service sector. Even well-intended policies may inadvertently create a negative impact if they are implemented inappropriately or are done so without understanding the context in which they are being implemented. Policy influences the sustainability of operations carried out by financial service providers. This influence can be direct through laws that inhibit certain actions or through practices that promote a decline of the sector. Some recent policies in Zimbabwe that have affected economic affairs include the Presidential Announcement of 7 May 2022 (GOZ, 2022) implemented to allegedly reduce the freefall of the Zimbabwe dollar against other currencies by suspending borrowing facilities which were "being abused by banks". The effect of removing the ability to borrow or lend prevented businesses and individuals from accessing liquidity for their daily operations. The outcry to this policy was immediate, as large businesses overnight struggled to continue operations. A second example is Statutory Instrument (SI) 213/2019 or Amendment of Exchange Control Regulations-2019 (GOZ, 2019b) and SI 212/2019 or Exclusive Use of the Zimbabwe Dollar for Domestic Transactions-2019 (GOZ, 2019a) where the GoZ outlawed the use of the United States dollar as legal tender. Despite this announcement, the same government would in a similar vein demand taxes and duties in foreign currencies creating confusion and sentiments that double standards are applied for self-serving government interests.

Legislation can, therefore, be seen to create or remove barriers to financial inclusion as the negative impact of these two examples was felt almost immediately with significant negative consequences. It has been determined that financial constraints are more widespread in countries where judicial systems are broken or weak as stated by Miriu (2020). These weak institutions diminish the viability of financial institutions and by consequence diminish the impact of financial inclusion efforts.

According to the Centre for Financial Inclusion, Conroy et al. (2009) and Miriu (2020), an enabling policy environment is more beneficial than any direct government intervention through credit guarantee schemes, even if these schemes are promoting sustainable financing to BOP and other vulnerable communities. This is because the government alone cannot wholistically achieve any targets for equitable development. The development progress must be co-produced by citizens, development partners and the private sector. Miriu (2020) states that an uncertain policy environment is associated with a lower rate of inclusion because of lower potential for innovation and lower potential for the development of new banking infrastructure in a restricted environment.

Governments have a responsibility to address legal, policy and regulatory barriers preventing the development of microfinance and other forms of financing as indicated Conroy et al. (2009). Laws must seamlessly cover all aspects of the financial service ecosystems from product development, product promotion, protection (through intellectual property laws) and product provision through the protection of business interest. Business interests should be protecting both the providers of services and the consumers from unfair business practices.

There is a need to cater to migrant populations with policies that will accommodate the unclear status of the migrants without comprising the sovereignty of host countries. Conroy et al. (2009) argue that it is important to consider language needs and create opportunities for access including providing workers' remittances for people with no bank accounts and for those who do not have permission to be in the country or who are there for limited periods.

Importantly, for the development narrative of African policies, African governments should not "cut and paste" policies borrowed from more diverse economies, instead, they should be "best fit" for local contexts. Best fit implies they may have aspects of what other countries have implemented, but they are refined for local contexts and remodelled to cater to the needs and aspirations of local users.

Additionally, policies must be responsive to high changes in technology and refrain from inhibiting products that can promote inclusion. When frameworks for fintech and digital financial services are improved, they increase the confidence of financial service providers and increase their popularity and use by consumers. For example, cryptocurrencies are becoming a dominant trading vehicle, yet they remain largely unregulated in developing countries with either partial or complete bans in at least 41 developing countries, with some cryptocurrencies being subjected to anti-money laundering and anti-terrorist financing laws as stated by United Nations Conference on Trade and Development (UNCTAD, 2022). Their use needs to be investigated and supported with effective registration mechanisms, effective payment systems and appropriate advertising. Regardless, digital currencies are here to stay and must be governed well and equitably.

Although secondary to an enabling environment, government programmes or facilities can be beneficial to channel funds from centralized structures removing bottlenecks and giving space to smaller institutions and, therefore, improve access, for example, moving funds from commercial banks to microfinance institutions can be done to improve access an example used by Conroy et al. (2009) to show their role.

6.2.2 Institutions and Access

When considering access to financial products, the private sector has a significant role to play in promoting access through traditional as well as new and innovative technologies. The private sector is more agile and responsive to market changes and can, therefore, quickly respond with new products that meet market needs. The proximity of private sector players to consumers gives them a unique place in identifying and exploiting market shifts and consumer preferences. Private players can provide a more holistic service through commercial banks, agent banking, insurance and capital markets.

In terms of access, the state plays a secondary role. That role is still significant, however, as state institutions can be critical to roll out financial services and increase access through ministries and social welfare programmes. In this respect, the state can reach remote areas that private players may not consider viable enough to exploit, but the state can reach those communities simply because of its responsibility to citizens and the structures that exist within it. Governments can use their structures alongside other development partners to provide access through cash transfers or health insurance, provided to larger population groups as ways of promoting inclusive social policies.

In the current literature, however, the importance of informal or unregistered networks has not been fully explored. Naseer and Amir (2019) argue that ordinarily the informal sector is not well accounted for despite contributing 40–50% to the GDP of developing countries. It has been observed that 80% of MSMEs in Africa are from the informal sector, but their biggest obstacle is a lack of access to finance. Naseer and Amir (2019) go on to state that although microfinance institutions (MFIs) are a representative of the informal sector by becoming a vehicle of support, we are yet to understand how informal service providers impact inclusion. This could be in the form of illegal and unregistered operators who provide loans for communities at either very reasonable or very exorbitant fees. Despite their illegal nature, they are meeting a demand created by the inadequate nature of the current financial systems to meet the needs of underserved communities.

Partnerships created through capacity building can improve communication with various industry players and subsequently improve access. These partnerships can be between formal and informal structures. Regardless of their formality, cooperation can change access and must be prioritized between the public and private sectors.

6.2.3 Institutions and Technology

Conroy et al. (2009) explain that technology has played a pivotal role in creating systems and products that improve access. With time, these technologies have gone on to become avenues to improve access to other sub-sectors such as retail banking, wholesale banking, access to MFIs, franchise or agent banking. Technology has lowered transaction costs and managed to put banking needs into a handset at the simple cost of a data connection. Technology has made it easier for remote communities to access financial services, where this would have been previously impossible. Conroy et al. (2009) have gone on to argue that MFIs can provide backend technology to reduce the costs of delivery to consumers which is a big hindrance to inclusion. Technologies such as point of sale (POS) devices, plastic cards and mobile applications will allow the safe transmission of data in affordable ways.

6.2.4 Institutions Contributing to Financial Inclusion in Zimbabwe

Zimbabwe has several key institutions that have promoted inclusion. Formal institutions that operate in the financial services sector are banks (commercial banks, building societies and merchant banks), MFIs, mobile network operators, investment houses, capital markets, insurance agents and more recently fintechs, especially for remittances, money transfers and POS transactions. There are also social institutions which cannot be regulated, but their influence is nonetheless present, namely:

(a) Community associations (burial societies, women groups and youth groups).

(b) Social media communities which influence behaviours and how people receive products and services.
(c) Income saving and lending groups (ISALs) demonstrate how agency is created away from the eye of formal institutions to create savings and investment vehicles.
(d) Religious groups sometimes promote particular beliefs which may be contrary to or in support of inclusion or exclusion.
(e) Chiefs and village heads are responsible for the governance by and enforcement of customary tenure in Zimbabwe.

According to Chaora (2022), it is determined that MSMEs are predominantly using the following services to gain access to financial products and services:

1. Deposit protection.
2. Access to foreign currency.
3. SME advisory services.
4. Loans and equity.
5. Short-term financing for business projects.
6. Bank accounts.
7. Mobile money accounts.
8. Insurance products.
9. Investments and saving services.

6.3 The State of Inclusion in Zimbabwe

According to the Financial Inclusion Index developed by SIVIO Institute (2022), citizens represented by MSME score 56% rates of *exclusion* meaning only 44% are *included*. Men in Zimbabwe are significantly more included at 47% compared to women at 39%. Urban women have the same rate of inclusion as rural men at 42%. The order of inclusion for Zimbabwe is that older men are at the pinnacle of inclusion, followed by younger men, then older women and finally young women aged 35 and under. The data that is outlined in the following sections was derived from the SIVIO Institute Financial Index (2022) and the 2022 report by Chaora on the Financial Inclusion of MSMEs in Zimbabwe.

Mobile money access is strong even for women at 60%. Savings and investment access are also strong for both men and women. Investments and savings, however, are accessed through formal and informal institutions such as ISALs, forex purchases and buying property. Banking access is more prevalent in men than women. Largely, inclusion has improved, but much needs to be done in the sectors of insurance and account access as indicated in Table 6.1.

6.3.1 Bank Account Access

According to Chaora (2022), bank account access is available to an average of 36% of MSMEs. The top barrier to bank account access is poor security protocols pointing to a lack of trust which was found to be a significant barrier for 19.4% of our small enterprises who do not feel their money is safe in Zimbabwean banking institutions. High bank charges have been cited by 20% who have complained that bank services are unaffordable. Here we see that the institutions we build can be barriers if our pursuit for inclusion is affected by our need for huge profits or where inflation promotes a speculative culture amongst service providers.

6.3.2 Mobile Money Technology

According to Zimstat (2022), mobile penetration rates are exceptionally high in Zimbabwe standing at 87%. Mobile phone operators have, therefore by default, become significant players to promote inclusion with Karombo (2022) pointing out that mobile money accounts currently stand at 4.1 million subscribers in Zimbabwe. The data gathered for the study by Chaora (2022) shows that 58% of MSMEs have access to mobile money accounts, but up to 51% of those users complain of withdrawals and transfer restrictions as a barrier to use on mobile money platforms. In addition, close to 30% of mobile phone users have complained of charges being excessive. The withdrawal and transaction limits are put in place to curb the operations of the parallel market

Table 6.1 Sector status of financial inclusion in Zimbabwe

Area of access	Men	Women
Banking and account access	44%	25%
Insurance access	24%	12%
Savings and investment access	50%	48%
Mobile money access	52%	60%
Overall inclusion score for Zimbabwe	**47**	**39**

Source: SIVIO Institute Financial Inclusion Index (SI-FINDEX) (2022)
NB: Other industry sectors have been excluded from Table 6.1; hence, direct averages do not apply to the overall inclusion score.

where transactions of USD to ZWL are done at rates far above the official rate governed by the RBZ. Whilst the charges include regular bank charges as well as the 2% Intermediated Money Transfer Tax implemented in 2018 (Ministry of Finance, 2018) by RBZ Governor Professor Mthuli Ncube as stated by the Ministry of Finance and Economic Development in Zimbabwe. To curb the parallel market, the RBZ put in place transfer restrictions to increase the charges associated with illegal trading, weaken the lustre of parallel market, and tax the illusive informal sector whilst simultaneously raising money for the government. If mobile money is to truly continue its usefulness in an inflationary Zimbabwe, then the Government should reconsider the limits regularly imposed on digital transactions and focus instead on ensuring that it deals with digital crime and economic recovery so that polices become regulatory rather than authoritarian.

We know that according to Chaora's report (2022) on MSMEs, at least 10% of MSMEs have complained of transactions being lost or unaccounted for, resulting in a mistrust of the financial system. Therefore, trust in the system becomes important. Not only is trust in the financial institutions important, but confidence in the government's reliability when dealing with pressure during a recession is also important. Zimbabwe has gone through successive periods of financial losses due to inflation and poor policies where people have lost savings overnight and enterprises have been reduced to bankruptcy because of abrupt policy changes. Although the desire to save and transact legally exists in the culture of Zimbabweans, the lack of trust is an important factor pushing people away from formal structures and legal transactions leading to savings, investment and forex purchases being done informally as noted by Chaora (2022).

6.3.3 Insurance Access

The Chaora (2022) report on financial inclusion by MSMEs shows that insurance products are used by 25% of men- and 15% of women-led MSMEs. Typical products used by MSMEs include personal health insurance, vehicle, staff, product, non-comprehensive business and comprehensive business insurance. The most accessed insurance product is vehicle insurance. However, the unaffordability of insurance products is a concern for up to 25% of insurance users, and up to 15% do not see the value of insurance products altogether. This points to the failure of institutions to make insurance products in ways that fit best for consumers, especially for rural and urban women and youth. Insurance products are designed in a generic manner, but the needs of women and youth are not generic. Rural women and urban women are different, and therefore their insurance needs are different. The needs of youth are also dynamic and wide-ranging depending on their education, race and current occupation. In addition, insurance companies need to understand that up to 20% of MSMEs are complaining about the impact of inflation stating that inflationary depreciation makes the products irrelevant, and MSMEs, therefore, no longer find any value in using them.

6.4 Microfinance and Third-Party Providers

MFIs face stringent registration requirements including the need for annual registration. Although these requirements have been put in place to regularise the sector and prevent consumer abuse, they have also served to create restrictions on lending which reduces the impact of MFIs to promote inclusion. Our approach, therefore, with MFIs should be co-creational that seeks to create a mutual understanding of the challenges and opportunities from the state and from the MFIs in Zimbabwe. The interventions must be co-creational to better understand the restrictions around microfinance with the aim of lowering transaction costs and, therefore, interest rates for BOP communities.

6.5 Social Institutions

Social institutions that are prominent in Zimbabwe include religious organisations, institution of marriage, and loose communities of individuals affiliated to community associations such as burial societies, ISALs and more recently online communities. Jowah (2020) demonstrated the role faith-based movements play in promoting giving and generosity in communities in Zimbabwe. We have also witnessed giving which is mandated as a percentage of income in Christian or Islamic religions amongst others. We have seen how these groupings can promote the support of new businesses. But we have also seen how religion can impede the access or productivity of women by encouraging them to stay home, restricting their access to money, and encouraging patriarchal access to finances or naming of assets. Savings and investment decisions in business were found to be delayed or deferred where no agreement was found especially where women entrepreneurs were concerned, implying they could not move forward without the consent of a spouse. This is evidenced by 69% of women-led businesses not having investments of any kind as stated by Chaora (2022). A top reason cited by Chaora (2022) was the lack of agreement with partners before investment decisions are made.

6.6 What Should Institutions Do to Promote Inclusion?

Registered institutions need to do more externally and with their clients to understand client-segmented perspectives. This will involve a commitment to understanding the challenges communities face in their interaction with various financial products and services. The product offerings in Zimbabwe should be modified with more consultation. There is more that can be done for young women in both rural and urban Zimbabwe using existing institutions such as schools or colleges to promote an understanding of financial products and services. Education needs to move beyond arithmetic and language to include an understanding of savings, investments and basic products and services that can promote an entrepreneurial and financial culture for MSMEs and citizens.

More needs to be done for women to ensure the products fit into what they need. We need to take time to find out what products they need and how they are using them including rethinking products such as health and business insurance, micro insurance and different forms of agent access to bring them closer to a customer-focused design form and function. At a social level, the social capital should be understood to have tailor-made products for women in different areas noting, of course, that urban women will have different needs to rural women and urban women will have different needs depending on whether they are in townships or suburbs. Research is imperative to understand the institutions surrounding financial access. There is a need to work with them to co-create local solutions that are best for local audiences and vulnerable BOP communities. Our research can be used to rethink credit, investment and savings facilities. Products and services must provide value that will meet consumers at their point of need.

With the significant role that technology plays in running payment gateways, operating banking systems and creating access to the Internet through mobile or other devices, access to energy is increasingly becoming significant for true financial inclusion to be possible. The cost of access to energy and Internet data will determine the level of inclusion in the immediate future in addition to long-term impact of all the inclusion efforts that have been carried out to date.

Our know-your-customer (KYC) requirements should be simplified and non-cumbersome, especially for women and youth in business, taking advantage of unique citizen identifiers to avoid fraudulent behaviour and excessive debt or illegal activities.

Formal institutions including the government and private sector players must understand the role that trust plays in product uptake and focus on consumer protection and consumer rights. If these are strengthened and mechanisms put in place to ensure consumers do not lose their savings, investments or bank balances when there is an economic downturn, this will improve significantly the rate of uptake. Trust in the system is important and, therefore by default, consumer protection becomes very important since Zimbabweans have survived several periods of loss due to recessions since the year 2000. During these recessions, banks largely made no effort to pay for losses incurred by citizens but rather prioritised their survival as institutions, eroding consumer trust.

Our support of fintech needs to grow, and so too must our emphasis on the digital landscape and its potential future role to deepen access. Our future is digital, and we need to start innovating now with BOP communities in mind in order for our products and services to be responsive and relevant to future needs.

The primary focus of all institutions should be promoting efficiency, quality and reducing costs. This can be done directly through adopting or developing systems that are more efficient. It can also be done by focusing on economic recovery. Several of the challenges or barriers to access stem from a system built on an unstable economy. If the economy is in a better position, then inflation would slow down, interest rates would decline, financial institutions would be in a better place to invest in systematic and human resource improvements and the financial service sector would likely perform better. A focus, therefore, must be on dealing with and removing economic crime followed by a systematic and direct focus on the state of the country's economic performance.

Over and above the initiatives developed to date, Zimbabwe needs to produce sound financial development data to provide information on her nuanced relationship between citizens and financial institutions. This will include creating avenues within all the private and public institutions to measure the access, quality, level of usage and effect of the use on the welfare of users. Of particular importance is the need for financial products and services to be tailor-made to local consumers and locally identified vulnerable groups. If such information is collected at a statewide level, not only will our policies be better directed, but the private institutions partnering with the GoZ will have more information on how the experiences of citizens can be deepened to make inclusion efforts more meaningful to Zimbabwean citizens. The definition of inclusion posited by the Reserve Bank of Zimbabwe measures effective usage, but it does not measure the quality, or barriers to product use, nor does it take into account the transformational capacities or lack thereof of financial products and services. According to the Alliance for Financial Inclusion (2010), the current data may be susceptible to double counting, for example, counting loans to women and loans to MSMEs twice. The data we have may not adequately capture the groups that are not being served by our financial inclusion efforts. If communities are being served, the data from groups serving communities in off-grid locations still remains hidden despite its valuable potential to the country's development narrative.

Zimbabwe's new financial inclusion strategy currently being drafted should consider and add to a more nuanced plan of implementation with a particular focus on financing for livelihoods in efforts to achieve middle income status by 2030 as stated in Vision 2030.

Conroy et al. (2009) explain that some key policy areas for African governments (including the Zimbabwean government) should be a focus on rethinking inclusion in the following areas: agent banking and their interactions with consumers, mobile banking, diversification of savings and insurance providers, reformation of public banking, development of financial identities for low-income clients and consumer protection in financial services. Policies in these areas can be revised to address the concerns related to improving information exchange employed by new or upcoming innovative technologies. In this interrogation, the best response options would be to close information gaps for all stakeholders in the face of rapidly changing technology and geopolitical events such as the recent Russia-Ukraine war and the COVID-19 pandemic. According to the Centre for Financial Inclusion (2019), governments need to develop and implement policies to address gender disparities in the quest for citizen access to key enablers such as mobile phones and national identity documents.

References

Alliance for Financial Inclusion (2010) Financial inclusion measurement for regulators-survey design and implementation

Centre for Financial Inclusion (2019) Global microscope 2019- the enabling environment for financial inclusion. Centre for Financial Inclusion

Chaora BR (2022) Financial inclusion of micro, small and medium enterprises (MSMEs) in Zimbabwe. SIVIO Institute, Harare

Conroy J, Ceasar PJ, Warapot M (eds) (2009) Promoting financial inclusion through innovative policies. Asian Development Bank Institute, Tokyo

Government of Zimbabwe (2018a) Transitional stabilisation plan 2018 to 2020. Government of Zimbabwe, Harare

Government of Zimbabwe (2018b) Vision 2030. Government of Zimbabwe, Harare

Government of Zimbabwe (2019a) SI 212/2019 exclusive use of the Zimbabwe Dollar for domestic transactions. Government of Zimbabwe, Harare

Government of Zimbabwe (2019b) SI 213/2019 amendment of exchange control regulations. Government of Zimbabwe, Harare

Government of Zimbabwe (2020) National development strategy 1- 2021 to 2025. Government of Zimbabwe, Harare

Government of Zimbabwe (2022) Measures to restore confidence, preserve value and restore macroeconomic stability. Government of Zimbabwe, Harare, p 1

IOSCO and OECD (2018). The Application of Behavioural Insights to Financial Literacy and Investor Education Programmes and Initiatives

Jowah E (2020) Perceptions of giving-findings from a survey on giving by Zimbabweans. SIVIO Institute, Harare

Karombo K (2022) Mobile money gains traction in Zimbabwe despite regulatory wrangling. https://itweb.africa/content/JN1gPvOYRYmMjL6m. Accessed Sept 2022

Ministry of Finance and Economic Development (2018) The two percent electronic transfer tax. Government of Zimbabwe, Harare

Miriu P (2020) The role of institutional quality in promoting financial inclusion. African Economic Research Consortium, Nairobi

Murisa T (2022) The past in the present: challenges of protecting customary tenure provisions- the Chilonga case. SIVIO Institute, Harare, p 2

Naseer I, Amir A (2019) Role of microfinance institutions in promoting financial inclusion and economic growth

Reserve Bank of Zimbabwe (2016) National financial inclusion strategy 2016 to 2020. Government of Zimbabwe, Harare, pp 1–15

SIVIO Institute (2022) SIVIO institute financial inclusion index. SIVIO Institute, Harare

Statcounter (2022) Statcounter. https://gs.statcounter.com/social-media-stats/all/zimbabwe. Accessed 28 Sept 2022

The World Bank (2022) GDP Growth (Annual %) - SubSaharan Africa, *The World Bank*, 4, CC-BY 4.0

UNCTAD (2022) All that glitters is not gold: the high cost of leaving cryptocurrencies unregulated. UNCTAD

Zimstat. Zimbabwe National Statistical Agency (2022). https://www.zimstat.co.zw. Accessed 29 Sept 2022

7 Fintech and Financial Inclusion: Closing the Gender Gap

Oluwaseyi Kehinde-Peters

Abstract

In sub-Saharan Africa, only 37% of women hold a bank account, compared to 48% of men, a disparity that has been exacerbated due to the COVID-19 pandemic. The situation is considerably worse in North Africa, where around two thirds of the adult population is unbanked, and the gender gap in access to financial services is the highest in the world at 18%. This chapter explores the identified barriers both on the demand and supply side and details the available fintech solutions, the various technologies underpinning these solutions, as well as case studies of how these solutions can significantly bridge the financial inclusion gaps for women in Africa with a view to spark ideas, innovation, and initiatives that will empower African women and eliminate barriers to financial inclusion.

Keywords

Fintech · Financial inclusion · Gender gap · Fintech solutions · Innovation · Africa

O. Kehinde-Peters (✉)
Pan African Women Empowerment Network (PAWEN), Nottingham, UK
e-mail: oluwaseyi.kp@pawen.org

7.1 Overview of Fintech and Financial Inclusion

Financial inclusion lays the foundation for a prosperous and sustainable economy. The United Nations identifies financial inclusion as an enabler for 7 of its 17 Sustainable Development Goals – SDGs (The World Bank 2022a, b). Additionally, the World Bank considers financial inclusion a practical approach to reducing poverty and enhancing shared prosperity, especially in developing countries, such as sub-Saharan African (SSA) countries (Omar and Inaba 2020).

Financial inclusion has been measured through the narrow lens of banked versus unbanked. Therefore, it is easy to assume that someone with a bank account can automatically access financial services and products necessary to meet their needs. However, this is a limited view that does not help achieve the ultimate goal of financial inclusion – eradication of poverty, which is the first SDG.

Financial inclusion means the availability of useful financial products and services that can fulfil the users' needs – transactions, payments, credit, savings, and insurance – delivered responsibly and sustainably (Mhlanga 2022). According to the Global Partnership for Financial Inclusion (GPFI), there are three dimensions in measuring financial inclusion: (1) access to financial services, (2) usage of financial services, and (3) the quality of the products and the service delivery.

T. A. Ojo (ed.), *Women and Finance in Africa*, Sustainable Development Goals Series,
https://doi.org/10.1007/978-3-031-53337-2_7

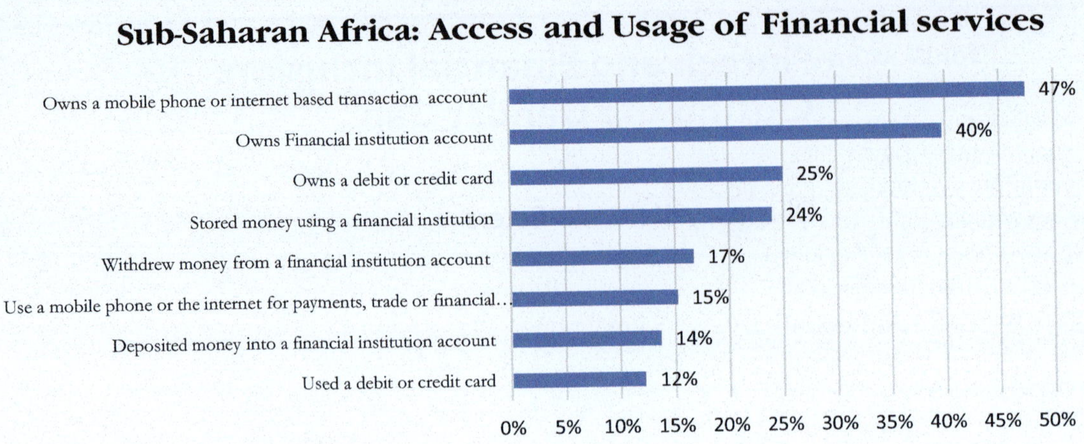

Data source: 2021 Global Findex report

The GPFI also has a basic set of indicators to measure financial inclusion, which are (1) formally banked adults, (2) adults with credit at regulated institutions, (3) formally banked enterprises, (4) enterprises with an outstanding loan or line of credit at regulated institutions, (5) points of service.

Access to transaction accounts – which serves as a gateway to financial services, enabling users to receive, send, and store money – is the first step towards accessing financial services. Financial access facilitates day-to-day living, helping individuals, families, and businesses plan for everything from long-term goals to dealing with emergencies. Accountholders can use other financial services such as insurance and credit to start and expand their businesses, manage risks, and invest in education and health, improving their overall quality of life.

According to the World Bank, remarkable overall progress has been made in the previous decades to achieve financial inclusion. It reported that the number of people with bank accounts increased by 1.2 billion between 2011 and 2017. The number of the unbanked population reduced further from 1.7 billion in 2017 to 1.4 billion in 2021 (The World Bank 2021). Despite the apparent progress, significant challenges remain.

First, women's access to financial services through formal systems such as bank accounts remains disproportionally low. The World Bank estimates that women are 6% more likely to be unbanked than men. However, this disparity has reduced from 9% in 2017.

Second, using the banked and unbanked parameters to understand the discrepancy in financial inclusion only tells half of the story. Merely owning a bank account does not automatically mean access to financial services and products. The graph below shows the disparity in sub-Saharan Africa between ownership and use of transaction accounts.

Therefore, our understanding of financial inclusion must take a more nuanced approach to fight poverty, especially the feminization of poverty in the continent. We must focus on access to finance and active use of financial services.

7.2 Women Financial Inclusion Gaps in Africa

The World Bank's *Global Findex Database* 2021 shows that developing economies contribute highly to the unbanked population. Africa is among the regions with the highest unbanked population – about 45% of adults do not have an account (Demirgüç-Kunt et al. 2022).

African women and men inclusive experience barriers to accessing financial services and products. Common challenges include insufficient documentation, cost of opening and maintaining bank accounts, physical access, and financial literacy. However, for women, these challenges are often more intense. For example, sub-Saharan Africa has a 15% gender gap in literacy rate, meaning that women face an additional literacy requirement barrier to accessing financial services (Wadhwa 2019). The barriers women face are two-pronged: the demand side – barriers that make it difficult for women to access and use these financial instruments, and the supply side – the availability and accessibility of financial instruments that meet the needs of women.

Demand-Side Barriers

- *Income barrier*: This is a key demand-side barrier. The subsistent nature of earning and high poverty levels women face exclude them from accessing financial services. For example, there are 127 women for every 100 men aged 25 – 34 living in extreme poverty in SSA. The result is that women lack enough funds to open and maintain bank accounts.
- *Literacy barrier*: The 15% gender inequality in literacy rate, coupled with technology and financial literacy rate differences, further contributes to the financial inclusion gap. According to *The Global Findex Database 2021,* a third of mobile money account holders in sub-Saharan Africa need help to use their mobile money accounts, and women are 5% more likely to need help using their mobile money accounts. This affects their ability to access and use financial services and products.
- *Legal and cultural barriers*: Unfavourable regulations also play a role in sustaining the gender financial inclusion gap. Many African countries inherited legal codes from colonial powers, with provisions that discriminate against women, and have maintained these legal systems and gender biases, even post-independence. For example, the 1960 Spanish Civil Code requires women in Equatorial Guinea to get their husband's permission to sign a contract. Additionally, three quarters of economies in Francophone Africa have regulations restricting women's employment as per a 1954 ordinance from the former federation of French West Africa. Therefore, Niger, Chad, and Guinea Bissau countries require married women to seek their husbands' permission to open a bank account. Ironically, most European countries where these legal systems originated have weeded out the gender discriminatory elements from their codes and used law as a tool to enhance gender equality (Christopherson et al. 2022). Besides regulations, many African cultures also work against women regarding financial inclusion. In Kenya, for example, the constitution grants women and men the equal right to property access; however, some tribes and religious groups do not abide by this rule and instead apply their own customary laws. In such communities, women are restrained from accessing properties and exploring other jobs and activities that would improve their financial status. Such gender stereotyping creates difficulty for women to access credit, as lending institutions may perceive them as high-risk candidates. Such limitations may be difficult to deal with as they are deeply embedded in society.
- *Trust and limiting-belief barrier*: Women disproportionately lack trust in banks compared to men usually based on negative past personal and narrated experiences. Some of these experiences have also been internalized, contributing to perceived low creditworthiness. A study done in the credit market of 47 countries shows that more women entrepreneurs in Africa are more likely to self-select out of the credit market due to low perceived creditworthiness (Morsy et al. 2019).

Supply-Side Barriers

- *Mobility constraints*: Some women find it more difficult to travel to the bank to access financial services and products due to factors such as household and childcare responsibilities, security, and cost. Cultural mobility issues make solo travel for women more difficult (Spencer et al. 2018).
- *Lack of identification documents*: In many places, women are disproportionally unlikely

to have identification documents compared to men. According to the recent *Women, Business, and Law Data*, women still face legal obstacles when applying for national ID cards in nine economies. For instance, in Benin and Namibia, a married woman is required to present a copy of their marriage certificate to apply for an ID card, yet the same does not apply to married men. Additionally, the sub-Saharan region has the largest proportion of adults lacking ID (30%) (The World Bank 2017).

- *Insufficient collateral*: The proof of collateral is always a minimum requirement for credit products. However, due to legal, economic, and cultural factors, women in some economies do not have equal opportunities as men to have the necessary collateral to get credits. For example, in Kenya, women only hold less than 10% of the registered title deeds, which is common collateral for issuing loans (Wasunna-Ochwa 2021).
- *Inappropriate set of product offerings*: Many providers just put pink in the marketing campaign to "incorporate gender" into products whose design fails to consider the specific needs and preferences of women (Gammage et al. 2017).

7.3 Consequences of the Gender Gap

The gender gap in access to finance significantly impacts social and economic progress in Africa, as women comprise 50.2% of the sub-Saharan population (The World Bank 2022a, b). This is half of the continent which cannot be financially excluded if the continent is to achieve its economic development goals. Today, women dominate some of the most critical sectors of the African economy, such as agriculture (Morsy et al. 2019). However, without access to financial services, such women cannot invest in modern technology to help them make more, manage, or multiply money. For example, when a woman farmer lacks financial services, she cannot diversify her farm, grow high-value crops, or invest in assets such as high-quality livestock. Additionally, the fashion industry in sub-Saharan Africa is dominated by women-owned small- and medium-sized enterprises (SMEs), a vibrant sector that commands $31 billion in annual revenue. However, women struggle to tap into the higher-value areas of the industry due to a lack of access to financing.

7.4 Addressing the Gendered Financial Exclusion Through Fintech Solutions

The solution to the gender gap in access to finance in Africa must consider the three dimensions of financial inclusion – access, usage, and quality of financial services, all of which financial technology (fintech) solutions provide. Fintech is defined broadly as encompassing advances in technology and changes in business models that have the potential to transform the provision of financial services through the development of innovative instruments, channels, and systems (Cambridge Centre for Alternative Finance 2020a, b). It applies to any new technology and innovation that competes with traditional economic methods to deliver financial services in the context of the three dimensions of financial inclusion. Fintech solutions are more agile, targeting underserved populations or providing better services and products. You can think of the difference between walking into a bank to request your balance and the ability to access the information in real time on your smartphone.

Fintech covers a wide range of financial solutions, from consumers' ability to track their transactions to applications that enable users to make local and cross-border payments to tools that analyse transactions and payment histories to ascertain credit ratings.

Fintech is transforming the access to, usage, and quality of financial products in unprecedented ways. It is destroying the old silos, integrating systems, and improving the consumers' experience with financial services and products. Over the last decade, 1.2 billion adults who were previously unbanked gained access to financial services through fintech (Appaya 2021). Fintech continues to help make financial services more accessible to an increasing audience.

7.5 Key Fintech Categories

According to the Cambridge Centre for Alternative Finance (CCAF) (2020a, b), fintech companies broadly fall under any of the following 11 verticals as shown in Fig. 7.1.

7.5.1 Digital Banks

The variety of fintech technologies and innovations has triggered digital transformation even in the most conservative traditional financial institutions. Many banking institutions have developed web and mobile applications to make their products and services more accessible to their customers. However, having mobile applications for well-established banks is just one aspect of digital banking. Recently, there has been a rise in digital banks that have only a virtual presence.

South Africa–based TymeBank is one of the largest branchless banks in Africa. The institution relies on its Android banking app and the official website to deliver its services to its clients any-

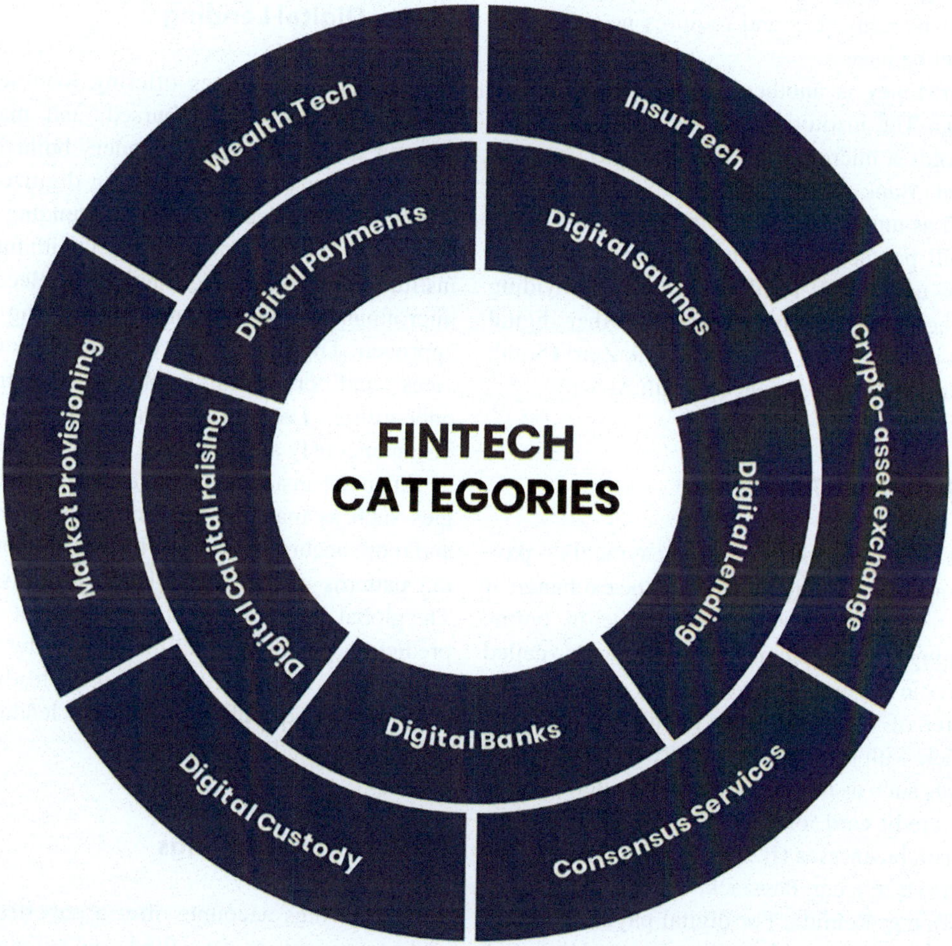

Fig. 7.1 Fintech categories. (Data Source Cambridge Centre for Alternative Finance (CCAF) 2020a, b)

where, anytime. Users can also access bank services through cell phones with a simple menu that makes sending money and checking their account balances easy (TymeBank 2023). Cell phone banking that supports USSD means that users who do not have access to the Internet or smartphones can still access the institution's banking services. TymeBank also collaborates with Pick 'n Pay and Boxer to facilitate easy account opening through self-service kiosk functions in the two retailers. Additionally, it allows users to bank at the tills where they shop. Another major selling point of TymeBank is the low cost of opening and maintaining accounts. By not having branches, the institution cuts operating costs which enables it to offer financial services to customers at a significantly reduced price. For example, it charges no fees for opening accounts and no monthly fees and requires no minimum account balance.

FairMoney is another digital bank based in Nigeria. The institution operates credit services and holds a microfinance bank license from the Central Bank of Nigeria (CBN). FairMoney started as an online lender offering instant loans and bill payments. Currently, it allows users to borrow up to NGN 1 million. It is also providing bank accounts with free transfers. Other digital banks include Kuda (Nigeria), Bank Zero (South Africa), and Better Finance (South Africa).

7.5.2 Digital Payments

There has been a rise in digital payments – payment methods that do not involve the exchange of cash – especially with the increase in online shopping. Such payments, sometimes called electronic payments (e-payments), consist of the transfer of value from one payment account to another, with the transacting parties using digital devices such as mobile phone, computer, prepaid card, credit card, or debit card. These could be between businesses (B2B), e.g. Duplo, with customers (B2C), e.g. Paystack, or with government (B2G), e.g. Remitta. For digital payment to take place, the payer and the payee must have an online banking method, a medium of transmission, and a device for the transaction. Therefore, users must sign up with a payment provider or an intermediary such as a service provider or bank.

Cellulant is one of the leading digital payment solutions in Africa. The company and its one-stop shop payments platform, Tingg, provide online payment solutions to consumers, banks, and consumers. It was recognized as Africa's leading digital payment solution provider at the World Business Outlook Award 2022. Cellulant acts as a bridge between financial institutions such as banks and users, enabling easy management of financial accounts and transfer of funds through mobile banking, messaging services, and digital channels such as mobile apps and USSD (Cellulant 2023).

7.5.3 Digital Lending

Digital lending involves offering loan services that are applied for, disbursed, and managed through digital channels. Lenders build intelligent customer engagement, using digitized data to inform credit decisions. Digital lending is disrupting the credit market in Africa, with financial institutions such as banks, neobanks, telecos, and microfinance institutions embracing this approach. Digital lending cuts the cost of services and enhances reach by streamlining onboarding. Lenders can also approve loans instantaneously and remotely.

Financial institutions use emerging technologies such as machine learning to make quicker and more accurate decisions by focusing on lending patterns and trends rather than individuals. The global market for digital lending platforms is predicted to reach 20 billion by 2026 (GlobeNewswire 2023). Some of the leading fintech companies offering digital lending are OneFi, Tala, and Branch.

7.5.4 Digital Savings

Digital savings accounts offer a paperless, fast, and secure way to save funds on a digital platform. These accounts support convenient ways to deposit and manage savings, such as instant transfer, SMS banking, and phone banking. Most digital savings accounts have very low to no minimum savings limits. Therefore, they encourage African women with a scarce and irregular income stream to save. It also breaks the mobility barrier, mainly affecting African women, by offering secure and low-cost solutions accessible through mobile phones and retail agents. M-Shwari, a combined savings and loan product launched by the Commercial Bank of Africa (CBA) in collaboration with Safaricom is one of Kenya‘s leading digital lending platforms. It has helped many women to get instant loans whose approval is based on digital data.

7.5.5 Digital Capital Raising

Fintech has given rise to new ways of raising capital. It offers ways to mobilize financial resources for start-ups without turning to traditional financial institutions. Even small enterprises can now find investors through various fintech services.

Digital capital enables customers access to investments and raises capital in exchange for equity or interest. Crowdfunding is the most famous alternative way of financing. It helps entrepreneurs fund and markets their businesses directly to the target audience. All that is needed for successful crowdfunding is an online platform, a good pitch, and attractive perks for investors. Thundafund (South Africa), M-Changa (Kenya), and UpriseAfrica are examples of crowdfunding platforms in Africa.

7.5.6 Digital Custody

Digital/crypto assets custody is a broad term that covers various ways of storing and protecting digital assets on behalf of the owner. The need for digital custody has grown with the rise in digital assets (cryptocurrencies) in the recent past. Custodial digital asset services require specialized technology that combines security, operational flexibility, and scalability. The availability of digital custody is significant to the continuing expansion of crypto assets and all other digital assets. Cold (offline) and hot (online) cryptocurrency wallets help users store and manage their digital assets. Examples of such wallets are Gemini, Ledger, and Coinbase wallet.

7.5.7 Consensus Services

As more organizations and businesses continue to use blockchain-powered financial applications, the demand for consensus-as-a-service continues to grow. Financial services that want to preserve data privacy while benefiting from decentralization can use consensus algorithms to improve trust. These algorithms enable users to facilitate and approve transactions in a decentralized and secure manner. Different blockchain networks adopt other consensus mechanisms. For example, Bitcoin and Ethereum 1.0 use proof of work, which essentially means doing some work to earn the chance to confirm a transaction. The process is transparent, as any user with the needed computing resources can take up the task. Other blockchain networks use different consensus algorithms, such as proof of stake, which is considered faster, cheaper, and more efficient.

7.5.8 Crypto Asset Exchange

We cannot talk about cryptocurrency as revolutionary assets in the finance sector without mentioning crypto exchanges. Unlike fiat currencies, crypto assets cannot be exchanged or traded in banks or conventional financial markets. This calls for platforms suited explicitly for these assets – digital asset exchanges. Crypto asset exchanges are digital platforms that facilitate swapping, buying, and selling digital assets. They are accessible from anywhere by anyone with a connection to the Internet. Binance and Coinbase are some of the leading cryptocurrency exchanges.

7.5.9 InsurTech

Fintech solutions have also impacted the insurance sector. Customers' demand for an effortless way of contacting their insurance companies and comparing offers from different providers without leaving their homes is a rise in demand. InsurTech is a technological innovation aiming to improve the insurance industry's efficiency. Fintech solutions touch consumer convenience and experience by enabling customer-based, flexible solutions for insurance services and products.

Insurtech is being developed to help those who cannot access insurance through conventional insurance services and products. For example, the award-winning Kenya-based PULA helps people, particularly smallholder farmers, access insurance quickly. It partnered, in 2019, with the World Food Program to insure 3.5 million farmers across ten countries in Africa (Santosdiaz 2021).

7.5.10 Market Provisioning

Market provisioning is a fintech area that uses smarter and faster machines to facilitate trades. It uses powerful engines to analyse big data to create trends and suggest trades. For example, they can read the news and people's actions in social forums to predict price changes and potential gains in investments. It relies on artificial intelligence (AI) and machine learning (ML) to run predictive modelling and self-correcting to improve accuracy, consistency, and trading speed. Some of the big players in this field are SAS (big data), SemLab (machine-accessible data), and Sentient technology (AI and ML).

7.5.11 Wealth Tech

Wealth Tech is a field that creates digital solutions to transform the investment and asset management industry. This vertical of fintech offers wealth management advice based on AI and big data, micro-investment platforms, and social network–based trading solutions. Wealth Tech involves micro-investments, which enable users to invest limited amounts with zero commission fees. Another niche in this field is Root-advisors, which create viable investment options through machine learning algorithms. Digital brokers also fall under this category, and they involve making stock data and opportunities available to people. An example is eToro. Other subcategories include investment tools and portfolio management.

7.6 Key Technologies in Fintech

Technological progress and innovation underpin fintech but vary considerably. Each use case may be unique, but the underlying theme is a common effort to open the financial sector to users of all places, classes, and genders. Here are the significant technologies driving the fintech business models.

7.6.1 Mobile Banking

Mobile banking uses smartphones, feature phones, or other cellular devices to access and operate bank services and products. Mobile banking can use a mobile application or unstructured supplementary service data (USSD) code to connect users and the benefits. The banking tasks that can be operated via mobile banking include account opening, checking account balance, requesting statements, transferring funds from one account to another, applying and repayment of loans, and locating the nearest bank branch or ATM. Some leading names in Africa when it comes to mobile banking are M-Pesa (Kenya), MobiKash (Kenya), Momo (Nigeria), Ecocash (Zimbabwe), and mTopUp, which is present in Egypt, Tunisia, Libya, Morocco, and Algeria.

7.6.2 Application Programming Interface (API)

Application programming interface (API) traditionally refers to the tech interface that sits between two software programs. API facilitates third-party applications in the banking system and connects to a bank's services and tools.

API banking involves protocols that avail bank services to other third-party companies through APIs. Therefore, it enables banks and third-party companies to combine their complementary specialities to offer better services to their customers.

APIs are significant to fintech companies and banks, providing better means to share data, personalize services, integrate with systems, and make financial services fast and efficient.

Banking-as-a-service (BaaS) is one of the applications of API. BaaS involves building a digital banking experience for customers, including online banking and mobile banking. Another API application is embedded finance, which embeds banking or financial tools and services in non-financial apps. Such devices and services include payments and lending. Okra, Mono, Dtitch, OnePipe, and Bloc are famous players in this field.

7.6.3 Blockchain

In this list, blockchain might be the most difficult to understand and implement, perhaps because it is one of the most recent revolutionary technologies in the fintech space. Blockchain is a distributed ledger technology (DLT), meaning it shares access to all the databases in the network among all users. Blockchain stores information electronically in a digital format. Blockchains are famous for their role in cryptocurrency, where they help record transaction data in a transparent and immutable manner. However, the technology can be deployed in many other areas in fintech. Blockchain technology guarantees fidelity and security of data records without the need for a trusted third party.

Blockchain technology is already disrupting the financial sector through its core feature of decentralization. Additionally, it powers financial products such as cryptocurrency, non-fungible tokens (NFTs), and decentralized finance (DeFi), creating a financially inclusive environment.

Interested parties are working to leverage blockchain technology to create financially inclusive products and services. For example, organizations are exploring the actual use cases of blockchain in creating a financially inclusive environment. For example, in Kenya, United Nations Children's Fund (UNICEF) spearheaded Kotani Pay's development, which allows users to convert cryptocurrency to Kenya shillings through their mobile phones (UNICEF Venture Fund 2021).

7.6.4 Artificial Intelligence

Artificial intelligence (AI) involves computers simulating human intelligence to mimic human behaviour in performing tasks. AI can repeatedly improve based on the information it collects, just as humans learn by practice and experience.

Financial companies can use AI to analyse and manage data from multiple sources to deliver essential insights. These innovative findings can help the bank improve its services, such as loan administration and payment processes. Additionally, AI powers many solutions to enhance security in the financial sector. For example, some banks use AI-powered biometric identities to allow users to access their services.

7.6.5 Big Data

The financial sector creates a large amount of data. Big data refers to a huge amount of structured and unstructured data that banks and financial institutions use to predict consumer behaviour and develop strategies to improve their services. Structured data exists within a company, and that is used to give insights into the crucial decision-making process. On the other hand, unstructured data are accumulated from various sources. They are rapidly increasing in number to provide considerable analytical potential.

Fintech companies use big data to predict clients' behaviour and develop risk evaluations. Such companies can react swiftly to the changing market. By adequately leveraging big data, fintech companies can understand consumers one-on-one, making better judgements and providing more personalized consumer experiences.

7.6.6 Decentralized Finance (DeFi)

DeFi is one of the financial sector's best disruptive applications of blockchain technology. Like conventional banking, DeFi allows users to access financial services such as lending, borrowing, fund transfers, trading, and savings. However, DeFi offers these financial facilities without intermediary entities such as banks. Instead, it relies on blockchain-based smart contracts. Smart contracts are computer codes with pre-programmed conditions which trigger an action upon meeting. DeFi products include non-custodial crypto wallets, decentralized exchanges (DEXes), and other financial decentralized applications (DApps) such as lending platforms. With no brokerage, exchange, or financial institutions involved, DeFi applications go beyond conventional boundaries to access regions, markets, and different layers of society.

7.6.7 Edge Computing

Edge computing is a new information technology (IT) architecture that distributes a range of networks and devices at or near the data source. This emerging computing paradigm enables processing of the user's data close to its origin. It enables faster processing of extensive data, leading to real-time action-led results. Edge computing allows fintech service providers to lower their cost of operation.

Edge computing can be employed in retail applications to analyse data generated from point-of-sale and merchandise-stoking operations and identify issues that need an immediate response. It also provides longer-term trends for sales and business opportunities.

7.7 Fintech Solutions Case Studies

7.7.1 Mobile Banking: M-Pesa

M-Pesa is a mobile phone–based banking product offering money transfers, payments, and micro-finance services. Launched in 2007 by Vodafone and Safaricom, a telco company in collaboration with the Commercial Bank of Africa (CBA), M-Pesa has grown to be the leading mobile money service in Africa. It boasts more than 604,000 active agents operating across seven African countries – Democratic Republic of Congo (DRC), Ghana, Egypt, Lesotho, Kenya, Tanzania, and Mozambique. M-Pesa has over 51 million customers who make over $314 billion in transactions annually (Vodafone Group 2023).

M-Pesa also runs a credit supply channel through commercial banks such as Kenya Commercial Bank (KCB) and savings and credit cooperative societies (SACCOs). M-Pesa also supports cross-border payments. M-Pesa collaborates with PayPal and an online payment system to enhance its operation, allowing users to receive foreign currencies and convert them into local currencies.

M-Pesa allows users to deposit money into their mobile-based accounts. They use personal identification numbers (PIN) or biometric identities to access and operate their accounts. Customers register at authorized agents and deposit cash in exchange for electronic money. They can send the money to other accounts by entering their PIN or using a biometric identity to confirm the transaction. They can also use the money in their accounts to pay for goods and services via *Paybill* or *Till* services. Additionally, users can deposit their funds in the virtual savings account, *M-Shwari*, which allows users to save funds and get loans. The collaboration of M-Pesa and commercial banks also enables users to transfer money between their bank accounts and M-Pesa accounts.

M-Pesa represents a significant improvement in national payment technology, making mobile banking easily accessible at a low cost. M-Pesa supports any mobile phone, including feature phones, thus increasing its accessibility. It also has a simple onboarding process and an easy-to-use menu that accommodates people with a low literacy level. As a result, it meets the needs of many women in small businesses who cannot use mainstream bank accounts due to low literacy.

M-Shwari uses customer data (M-Pesa usage) to determine users' credit automatically. Therefore, it bypasses the gender-based bias that banks may have on women when approving loan limits. Additionally, M-Shwari does not require any collateral. Therefore, women can get soft loans to boost their SMEs.

Users can deposit as low as one shilling to their M-Shwari, which encourages the saving culture for women with irregular income from their SMEs.

Financial institutions in Kenya have embraced M-Pesa as a platform for managing micro-accounts, building customer deposits, and expanding customer networks. Consequently, Kenya has become a leader in financial inclusion in sub-Saharan Africa. A decade ago, only 26.7% of the adult population in Kenya had access to formal financial services. By 2016, about 9 years after the launch of M-Pesa, the number rose to 75%. The number rose to 85% by 2021 (Njoroge et al., 2021).

Additionally, M-Pesa offers privacy, allowing women in patriarchal households to manage their liquidity with less control from male heads.

One of the challenges associated with M-Pesa is fraud. A common trick involves fraudsters calling or texting M-Pesa users, telling them they have mistakenly sent money to their accounts and requesting a refund. Many women have lost money through this means.

Another challenge relates to service for special populations. Elderly women may be unable to use their M-Pesa as they cannot read. Consequently, they end up sharing their PIN with others to help them complete transactions. Also, many members of the elderly women population may not have mobile phones, thus locking them from M-Pesa services.

7.7.2 Blockchain: The Grassroots Economics

Blockchain creates room for innovation in the financial sector. Grassroot Economics, a Kenyan-based project funded by UNICEF's Innovation Fund, targets to close the credit gap in low-income groups in the country. It involves using community inclusion currencies (CICs), tokens backed by actual goods and services in a specific community, such as water, food, or babysitters. People can use these tokens to secure lines of credits (Chapiro, 2021).

The CICs enable the monetization of resources, enabling users to give or take credit without the need for fiat money or banks.

Some women in the villages might not have a line of credit or a stack of cash to access credit. However, they have actual goods and services such as crops, clothes, or even the labour of schoolteachers or cooks. CIC holders can also get loans based on their historical trade – receiving up to 10% of their annual revenue as credit – allowing them to continue trading beyond their current liquidity.

Users can manage their tokens using cell phones, including feature phones. However, transactions are recorded in the blockchain, which offers a high level of transparency and security. This project already had 58,400 users with more than 500,000 transactions worth $three million (Railsr, 2023).

The Grassroot Economics project's main challenge relies on the novel blockchain and cryptocurrency technologies. Most people still do not understand how these technologies work and are unable or sceptical to use them. Additionally, it will reflect the gender gap in technology to gender gap access to finance.

7.7.3 API (Open Banking): Railsr

Formerly known as Railsbank, Railsr is a London-based open banking and compliance platform that provides digitally native companies with a connection to payment schemes and the ability to create bank accounts instantly. Railsr facilitates easy access to international banking.

One of the main products of Railsr is embedded banking and wallets – BaaS. It is a complete set of banking and wallet components that retailers can assemble to create any use case needing banking or wallet functionalities embedded in the customer experience. Such use cases may include holding balances, creating and operating bank account details, international transfers, FX, digital asset storage, bank account opening, sending and receiving funds, and inter-ledger transfers. The project also offers cards as a service (CaaS), which enables launching of an easy-to-use bank card system (virtual and physical cards) into customer experience and brand. Additionally, it offers embedded credit – credit as a service (CaaS), which includes credit application, balances and statements, and credit management.

Railsr is enhancing financial inclusion by enabling millions of users to upgrade to easy-to-use digital transactions in a secured environment. More than $2 billion of venture capital has been raised by projects supported by Railsr.

Banks and other African financial institutions can deploy Railsr services to reap benefits such as providing access to responsible credit, encouraging informed financial behaviours, and enabling participation in the global economy. The API services will enable easy onboarding, encouraging more women, especially those with low literacy levels. Railsr can also help banks and other financial institutions offer different global credit scoring benefits to women entrepreneurs.

Other services such as CaaS enable easy access to financial services, encouraging more women onboarding.

7.7.4 Artificial Intelligence: TymeBank

TymeBank is South Africa's digital bank that uses AI to interact with its customers online and through kiosks. The bank has fewer employees, 250 employees, compared to 50,000 employees in an average traditional South African Bank. Consequently, the bank adopts lower charges. For example, TymeBank's card replacement charges R40 compared to Standard Bank's R55 and FNB's R110.

Additionally, TymeBank is linked with the Department of Home Affairs database to capture biometrics for verifying customers' identities at kiosks. Customers can open a bank account at more than 800 stores that are allowed to install real-time biometric recognition kiosks. Users only need an identity number and no additional paperwork. A few minutes after registering, customers receive a Visa debit (Mambu, 2023).

TymeBank uses advanced modelling techniques with high accuracy to access loan payback probabilities. It enables risk-based pricing to determine customers' risks and price credit on an individual level, thus lowering credit costs and expanding the availability to a wider base. The model provides transparency in the credit.

Since its launch in 2018, the bank has seen tremendous growth, reaching 670,000 customers by August 2019 (Malinga, 2019).

7.7.5 Big Data: International Business Machines Corporation (IBM)

IBM is one of the biggest vendors of big data–related products and services. Its big data solutions cover data storage, management, and analysis. IBM is a dedicated cloud solution that offers deep customer insight in the form of a prediction about financial events, life events, the likelihood of buying or selling banking products, and even the propensity to leave the bank (C-Level, 2023).

IBM works with over 90 of the top 100 banks worldwide, leveraging big data to help the banks understand their customers' behaviour and deliver personalized banking experiences (IBM, 2023).

Through such technology, existing banks, financial institutions, and start-ups can understand women's behaviour in banking and meet their needs. For example, using big data provides a more reliable way of assessing an individual credit score in a non-biased way. Such a move will level the playground for women when taking loans.

7.7.6 Decentralized Finance: MakerDAO

MakerDAO describes its governance token DAI as "the world's first unbiased currency". MakerDAO is a DeFi protocol that allows users to lend and borrow funds. The protocol is run by a smart contract that ensures transparency and immutability of transactions. Users worldwide can borrow funds unanimously, as blockchain allows users to hide their identity of users. Therefore, it prevents any chance of discrimination when offering financial services. All the activities, including the lending and borrowing rates, are programmed. MakerDAO also supports savings, allowing users to lock their funds in the DAI savings rate (DSR) smart contract and earn interest that the token holders decide (Walker, 2021).

Registering with DAO is simple and does not require personal details. Users only need an account with one of the supported wallets to manage their digital tokens. The transaction fees for DeFi protocols such as MakerDAO are suitable for breaking the gender barrier in financial inclusion. First, it adopts a non-discriminatory way of handling the users. The system credit score is determined by smart contracts rather than manually, as with brick-and-mortar financial institutions. Consequently, both women and men receive equal treatment in terms of access to financial services and products. This feature is specifically important in Africa's financial inclusion context, where culturally informed gender-based biases still influence women's financial inclusion.

Additionally, it enables easy access to loans and lending services, as it does not require external collateral such as title deeds or Logboks, which most women in Africa lack. Instead, women can take loans based on their DAI savings.

The main challenge with DeFi solutions is the lack of liquidity. Most of the digital tokens have not broken into the mainstream. The technology is new and harbours novel risks, such as rug pulls. Therefore, most women may shy away from onboarding.

7.7.7 Edge Computing: Amazon Web Services (AWS)

AWS has invested heavily in edge computing, offering SMEs and large enterprises an extensive toolset. It provides cloud-edge services such as hybrid cloud, Internet of things (IoT), robotics, industrial machine learning, analytics, compute services, and industrial machine learning (ML). With quick and scalable applications, AWS boasts more than 200 integrated device services that clients can choose from (Amazon, 2023). It also provides connected vehicle solutions, AWS IoT camera connector, and IoT device simulator. AWS also sells its edge products – Alexa and Echo devices.

AWS serves financial works with institutions across banking, capital market, payments, and insurance to help offer quality services and products. Through low-latency analysis of large quantities of data, edge computing enables personalization of customer engagement down to the individual level.

Using this technology can help the bank understand the needs of women in banking at a more personal level. Therefore, they can create strategies to boost their services to the group. For example, by understanding their female audience, they can target them with appropriate advertisements and educative items to promote their engagement with the bank. Additionally, banks can use edge computing to create robot assistants that use natural language processing to interpret and comply with customer information requests. Through machine learning, such robots can go as far as perceiving human emotions and adjusting their behaviour to ensure proper interaction. Such assistants can be designed to help women access financial services and products easily.

Banks and financial institutions working with AWS include Barclays, Capital One, AON, Alliance, Stripe, and Robinhood (Amazon Web Services, 2023). The only challenge with AWS is that its interface is complex, which calls for companies that will handle their edge infrastructure to have considerable in-house expertise.

7.8 Conclusion

Women constitute about half of the African continent, and as such empowering women by closing the financial inclusion gap is a key imperative for the economic growth of the continent. Fintech has proven to have the solutions to address both the demand-side barriers and supply-side barriers to financial inclusion for women. On the demand side, digital payments enable women entrepreneurs to sell their products and services to a wider customer base through e-commerce and other related platforms, thus increasing their income. Digital banks, digital savings, digital lending, and digital capital–raising solutions make redundant the trust, legal, and cultural barriers to financial inclusion for women. On the supply side, digital banks eliminate mobility constraints and high transaction cost barriers. Digital lending and digital capital raising also reduce the collateral barriers women face. These solutions exist as seen in the highlighted case studies and have shown early successes in accelerating financial inclusion. However, there is a need for specific women-focused solutions that incorporate these features while addressing the financial literacy and lack of identification barriers women face. As the adoption of these solutions scales, concerted efforts need to be undertaken by the government and stakeholders to bridge the digital literacy gaps, as this in itself can create an unintended dimension of financial exclusion.

References

Amazon (2023) BBVA selects AWS to accelerate its data-driven transformation. Available at https://press.aboutamazon.com/2023/6/bbva-selects-aws-to-accelerate-its-data-driven-transformation. Accessed 20 Sept 2023

Amazon Web Services (2023) Financial services – AWS Cloud Governance for Financial Services. Available at https://aws.amazon.com/financial-services/security-compliance/cloud-governance/. Accessed 20 Sept 2023

Appaya S (2021) On fintech and financial inclusion. Private Sector Development Blog, The World Bank

Cellulant (2023) One-stop-shop payments platform for businesses across Africa. Available at https://www.cellulant.io/. Accessed 20 Sept 2023

Chapiro C (2021) Working toward financial inclusion with blockchain. Available at https://ssir.org/articles/entry/working_toward_financial_inclusion_with_blockchain

Christopherson K, Yiadom A, Johnson J, Fernando F, Yazid H, Thiemann C (2022) Tackling legal impediments to women's economic empowerment. International Monetary Fund

C-Level (2023) CLevel the Trusted Advisor. Banking and Financial Services. Available at https://www.the--c-level.com/industries-2/banking-financial-services/. Accessed 20 Sept 2023

Demirgüç-Kunt A, Klapper L, Singer D, Ansar S (2022) The Global Findex Database 2021: financial inclusion, digital payments, and resilience in the age of COVID-19. World Bank Publications

Gammage S, Kes A, Winograd L, Sultana N, Hiller S, Bourgault S (2017) Gender and digital financial inclusion: what do we know and what do we need to know. International Center for Research on Women (ICRW)

GlobeNewswire (2023) Digital Lending Market worth $20.5 billion by 2026, growing at a CAGR of 13.8% Report by MarketsandMarkets™. Available at https://www.globenewswire.com/news-release/2023/08/11/2723639/0/en/Digital-Lending-Market-worth-20-5-billion-by-2026-growing-at-a-CAGR-of-13-8-Report-by-MarketsandMarkets.html#:~:text=Chicago%2C%20Aug.,new%20report%20by%20MarketsandMarkets%E2%84%A2. Accessed 20 Sept 2023

IBM (2023) Unlock value with AI you can bank on. The future of banking is here. Available at https://www.ibm.com/industries/banking-financial-markets. Accessed 20 Sept 2023

Malinga S (2019) How AI helped TymeBank gain 670K customers. ITWeb. Retrieved from https://www.itweb.co.za/content/WnxpE74DX9K7V8XL

Mambu (2023) 'Mambu' South Africa's First Digital-Only Bank. Available at https://mambu.com/customer/tyme-bank. Accessed 20 Sept 2023

Mhlanga D (2022) The role of financial inclusion and FinTech in addressing climate-related challenges in the industry 4.0: lessons for sustainable development goals. Front Clim 4:949178

Morsy H, El-Shal A, Woldemichael A (2019) Women self-selection out of the credit market in Africa

Njoroge P, Obudho M, Cook T (2021) 2021 FinAccess Household Surveyl December 2021. Available at https://www.knbs.or.ke/wp-content/uploads/2021/12/2021-Finaccess-Household-Survey-Report.pdf. Accessed 20 Sept 2023

Omar MA, Inaba K (2020) Does financial inclusion reduce poverty and income inequality in developing countries? A panel data analysis. J Econ Struct 9(1):37

Railsr (2023) Railsr, Formerly Railsbank. The Global Embedded Finance Platform. Available at https://www.railsr.com/. Accessed 20 Sept 2023

Santosdiaz R (2021) The Fintech Times. Overview of Insurance and Insurtech in Africa Available at https://thefintechtimes.com/overview-of-insurance-and-insurtech-in-africa/. Accessed 20 Sept 2023

Spencer S, Nakhai M, Weinstock J (2018) The role of trust in increasing women's access to finance through digital technologies. USAID, Washington, DC

The Cambridge Centre for Alternative Finance (2020a) Cambridge Fintech Ecosystem Atlas. Available at https://ccaf.io/. Accessed 20 Sept 2023

The Cambridge Centre for Alternative Finance (2020b) New global COVID-19 FinTech study launches. Available at https://www.jbs.cam.ac.uk/2020/new-global-covid-19-fintech-study-launches/. Accessed 20 Sept 2023

The World Bank (2022a) 'Data',. Population, Female (% of Total Population) – Sub-Saharan Africa. Available at https://data.worldbank.org/indicator/SP.POP.TOTL.FE.ZS?end=2022&locations=ZG&start=1960&view=chart. Accessed 20 Sept 2023

The World Bank (2022b) Financial inclusion – Financial inclusion is a key enabler to reducing poverty and boosting prosperity. Available at http://www.worldbank.org/en/topic/financialinclusion/overview. Accessed 19 Sept 2023

The World Bank (2021) Global Findex – The Global Findex 2021. Available at http://www.worldbank.org/en/topic/financialinclusion/overview. Accessed 19 Sept 2023

The World Bank (2017) Global ID Coverage, Barriers, and Use by the Numbers: Insights from the ID4D-Findex Survey. Available at https://documents1.worldbank.org/curated/en/953621531854471275/Global-ID-Coverage-Barriers-and-Use-by-the-Numbers-Insights-from-the-ID4D-Findex-Survey.pdf. Accessed 19 Sept 2023

TymeBank (2023) 'Ways to Bank', Ways to Bank with TymeBank. Available at https://www.tymebank.co.za/ways-to-bank/. Accessed 20 Sept 2023

UNICEF Venture Fund (2021) Kotani Pay: Increasing Access to Financial Instruments Using Blockchain and Cryptocurrency. Available at https://www.unicefventurefund.org/story/kotani-pay-increasing-access-financial-instruments-using-blockchain-and-cryptocurrency. Accessed 20 Sept 2023

Vodafone Group (2023) M-PESA. What is M-PESA? Available at https://www.vodafone.com/about-vodafone/what-we-do/consumer-products-and-services/m-pesa. Accessed 20 Sept 2023

Wadhwa D (2019) More men than women are literate. Retrieved March, 13, p 2020

Walker A (2021) How to stake DAI. Earn interest from idle DAI tokens and expand your cryptocurrency portfolio. Available at https://www.finder.com/uk/how-to-stake-dai. Accessed 20 Sept 2023

Wasunna-Ochwa E (2021) While more women are starting and owning companies, financial inclusion remains low. Available at https://www.theeastafrican.co.ke/tea/oped/comment/-financial-inclusion-remains-low-for-women-3545272#:~:text=By%20ELIZABETH%20WASUNNA%2DOCHWA&text=In%20Africa%2C%20women%20are%20applying,women%20entrepreneurs%20in%20the%20world. Accessed 19 Sept 2023

World Bank (2021) World development report 2021: data for better lives

Financial Inclusion for Women in Nigeria: Much Ado About Access

8

Temitayo Adenike Adeyemo

Abstract

The case for financial inclusion in developing countries has taken center stage in development discourse in recent years, which may suggest significant improvement in financial inclusion in developing economies. Despite this, evidence still shows that financial inclusion is biased against women. The application of gender lens to the view of financial inclusion has usually looked at these differences between men and women via access to finance and credit facilities. However, the issues involved reflect both the supply side at the macro level and demand-side factors at the household level. It is thus pertinent to ask: to what extent is gendered budgeting reflective of financial inclusiveness in Nigeria? At what critical point does financial inclusion favor male over female? Do women within different spatial features have any edge in the financial inclusion game by any means? This chapter, therefore, examines financial inclusion from the multidimensional lens of opportunity, accessibility, and utilization for Nigerian women in general and rural women in particular. Findings reveal that there are demand and supply factors that lead to self-selection of women into degrees within the financial inclusion frame, with differentials across rural and urban sectors. Policy implications are likely to reflect the need to examine the phenomenon of financial inclusion beyond access to credit to a holistic view that includes creating a baseline opportunity to enhance women's capacity to access formal financial systems.

Keywords

Financial inclusion · Opportunity · Access · Women

8.1 Background

There has been an increased advocacy for financial inclusion in developing countries including Nigeria over the years. However, the gender disparity in financial inclusion seems to persist globally at 9% (Eckhoff et al. 2019) and up to 33% in Nigeria (Enhancing Financial Innovation and Access (EFInA) 2021). This may be related to a lack of understanding of underlying factors that could constrain the capacity of women to access and use financial services. Consequently, there are renewed calls for an adequate policy framework to develop more encompassing policies to improve financial inclusion.

This is on the premise that inclusion within financial systems has been reported as key to

T. A. Adeyemo (✉)
Department of Agricultural Economics, University of Ibadan, Ibadan, Nigeria

T. A. Ojo (ed.), *Women and Finance in Africa*, Sustainable Development Goals Series,
https://doi.org/10.1007/978-3-031-53337-2_8

many developmental goals on a global scale, with implications for poverty reduction, employment, prosperity, and well-being of people (Ouma et al. 2017; World Bank 2022). The idea of financial inclusion is such that it allows every facet of the economy, groups, households, and individual to have access to financial services available within their contexts and in a sustainable way. The World Bank has defined financial inclusion to mean a situation where individuals and businesses have uniform access to affordable and useful financial services (World Bank 2015; Kabakova and Plaksenkov 2018). Financial inclusion, therefore, drives economic development and seeks to integrate the informal and shadow economy for a more inclusive development (Kumar 2013; Younas et al. 2022).

Global developmental goals have pointed to the pivotal role of financial inclusion in sustainable development (World Bank 2022). Ensuring financial inclusion means that people are able to obtain and sustain a form of livelihood with consequences for poverty reduction and building resilience. This is because financial services are not limited to operating a bank account but also involve credit, insurance, and other financial transactions needed for people to have a seamless interaction within their societies (Ouma et al. 2017). Developed economies appear to have been able to ensure that a large proportion of their citizens are able to functionally participate in the financial markets. The participation of women in financial services has been shown to improve their livelihood and the welfare of their households through reduction of income inequality (Sarma and Pais 2011; Cabeza-García et al. 2019; Igwe and Adelusi 2021).

However, the extent of participation in the financial systems of developing economies is still subject to a series of bottlenecks that are disincentives to potential users. Even more affected are vulnerable groups, including women, youths, and disabled persons (Izquierdo and Tuesta 2015). This suggests such groups of people are also socially excluded with dire consequences for decision-making in many aspects of their living and livelihood formation (United Nations 2022). The Nigerian situation reveals a large number of unbanked adults, with many financial transactions occurring in the informal market in the country (Kama and Adigun 2013). Push by the global community of practice seeking to improve financial inclusion by at least 400 million people by 2020 seemed to have galvanized a number of economies to improve country-wide inclusivity (World Bank 2022) including Nigeria.

Financial inclusion in Nigeria seemed a mirage until policy measures to improve the coverage of financial services through the 2012 National Financial Inclusion Strategy (NFIS) to continually assess the extent of financial inclusion and develop a framework for reducing the level of financial exclusion in the country (Central Bank of Nigeria (CBN)/EFInA 2019). The expectation to have a form of improvement in the level of financial inclusion in general in the country was high; however, the basic measure of access to financial service seemed to have in reality witnessed a fall (Demirgüç-Kunt and Klapper 2012).

It is a worrisome issue, however, within the sociocultural context of the nation that despite the move towards financial inclusion, there seems to be a further deepening of inequality between men and women in the country. The financial exclusion figures of 24% and 36% for men and women, respectively, gives a gender gap of about 33% between male and female financial inclusion in Nigeria (Central Bank of Nigeria (CBN)/ EFInA 2019). This level of exclusion of women from the financial market has had grave implications for the livelihood structure of many women and households in Nigeria. This suggests that benefitting from the policy on financial inclusion is subject to a number of other factors other than the main components of availability, accessibility, and affordability of financial services. These other factors related to creating an agency for which women are able to get into the financial market appear to be the critical hurdle to be overcome for women's inclusion in the financial system in Nigeria. These societal norms and values segregate women within supposed "male-decision" processes in the country, limiting women to obtain productive livelihoods, participate in commodity markets, and become autonomous (Igwe and Adelusi 2021).

This chapter, therefore, postulates that it is important to examine other foci within the financial inclusion framework in order to have a holistic understanding of the concept and hence provide context specific recommendations as required. The aim is to understand the factors that interact to bring about demand for formal financial services as well as supply of these services to women. Therefore, this study asked: what are the opportunities that women have in getting access to financial services? And therefore, how accessible are financial services to women – urban and rural?

8.2 Historical Perspective on Financial Inclusion in Nigeria

The Nigerian government has described financial inclusion to mean a situation in which all adult Nigerians are able to gain access to a wide range of formal financial services at an affordable cost in order to meet their needs. In October 2012, the Central Bank of Nigeria set up the National Financial Inclusion Strategy, in collaboration with other stakeholders (Central Bank of Nigeria 2018), with the goal to ensure that all adult Nigerians could participate effectively in the financial markets. One of the vehicles of the NFIS was the Financial Inclusion Technical Committee, which was set up in February 2015 and tasked with the goal of increasing financial inclusion from the 2010 figure of 36.3% to 70% (formal financial sector) and 10% in the informal financial system by 2020. The intervention was to include such priority areas as simplification of the KYC system in deposit money bank system, implementation of regulatory framework for agent money banking, development of a National Financial Literacy Framework, mobile payment and cashless economy streams, and credit enhancement schemes. As of 2016, the percentage of adult Nigerians served by financial institutions was 58.4%, of which 33% was in the formal banking sector (Central Bank of Nigeria 2018).

A revision of the 2012 National Financial Inclusion Strategy was done between 2017 and 2018. The evaluation process engaged stakeholders in the private sector, public sector, and development partners, leading to a revision of the objectives of the initial National Financial Inclusion Strategy. The findings of the review revealed major pointers to build opportunities for the effective participation of adult Nigerians in the financial system. Some of the barriers identified include cultural and religious constraints, financial illiteracy, technology, as well as some transactional regulations (e.g. fixed transaction costs). The removal of these barriers would, therefore, constitute a baseline opportunity for financial inclusion in Nigeria.

As of 2020, the financial inclusion figure for Nigeria had increased to 64, with 45% in the formal banking sector (EFInA 2021). While there was an increase in financial inclusion, overall, there is still a large gap in meeting the target of formal financial inclusion in the country. In this empirical evidence, women have had lower access to financial services than men, and it is much worse for those in the rural areas (Central Bank of Nigeria (CBN)/EFInA 2019). However, there has been no specific framework to target financial inclusion of women in particular within the NFIS structure. The aim of this study is to generate empirical evidence to show underlying barriers to women's access to financial services in Nigeria. These barriers may be categorized as demand-side (women not making a demand) and supply-side (barriers to availability and ease of access) factors that need to be understood to create effective opportunities for women to access financial services in Nigeria.

8.3 Methodology

This study is a mixed method desk review of the financial inclusion situation of women in Nigeria from qualitative and quantitative perspectives. The qualitative method involved a desk review of the perspective of strategic policy documents and other literature related to financial inclusion and the opportunity dimensions that reflect access and utilization of financial services for women in Nigeria.

The quantitative component of this study explored data on financial inclusion from the Global Financial Index and the Enhancing Financial Innovation and Access (EFInA) platforms. The Global Financial Index (Demirgüç-Kunt et al. 2018) is a database of financial inclusion variables in 2011, 2014, and 2017 on a national level. The EFInA survey is a yearly survey on financial inclusion across various sectors in Nigeria. This study made use of the 2020 EFInA financial inclusion survey in its analysis to examine cross-sectional relationships between variables of financial inclusion and opportunity for financial inclusion in Nigeria.

8.4 Data and Data Analysis

The data needed for this study was mainly reflective of both the historical Global Financial Index and the cross-sectional EFInA data. The inclusion of some main social indices was also triangulated to understand the opportunity frame being discussed. Developing indices of financial inclusion involves an interaction of key components of access and utilization. In this study, these were identified as:

- Having a bank account.
- Having savings in a formal financial institution.
- Having access to credit.
- Having an insurance policy.

The analysis was descriptive and presented in tables with relevant illustrations.

8.5 Components of Financial Inclusion: How Does Opportunity Play a Role for Women?

Financial inclusion includes dimensions of access, quality, and usage, so inclusion occurs when an individual or a group has sufficient access to good quality financial service with evidence of reported use (World Bank 2015). However, the baseline factors that make women able to access and utilize financial services are usually absent in developing countries and more so in rural areas. Many studies have attempted to bunch these baseline premises as determinants or drivers of access to financial services (Govindapuram et al. 2022; Kaur and Kapuria 2020). In this study, we have termed these attempts as opportunities that women are able to harness to obtain and sustain financial services for their livelihood.

The argument for this is related to the ecosystem approach (Kabakova and Plaksenkov 2018; Tay et al. 2022), where financial inclusion was viewed via the lens of a series of baseline factors that must be put in place for financial inclusion polices to work. In their study, (Izquierdo and Tuesta 2015) Izquierdo and Tuesta explained these as critical factors for financial inclusion. Thus, the role of opportunity in driving financial inclusion reflects the need to create the baseline environment or ecosystem needed for strategies of financial inclusion to thrive. Financial inclusion will thrive in developing countries based on an integration of sociodemographic, sociocultural, economic, policy, and technological factors. These variables function as first-level determinants of inclusion or exclusion in the financial market for which policy seeks to drive equality of access.

The 2020 EFInA survey data helped to identify four main dimensions of financial inclusion having a bank account, savings in a formal bank account, access to credit, and an insurance policy. Other dimensions in the survey relate to pension fund and agency banking, but with little representation in the rural domain. Analysis of the 2020 EFInA data (Fig. 8.1) shows that 72% of adult women in Nigeria have a bank account, while 24% use these formal channels for savings. Less than 5% of women have access to credit facilities (3%) and an insurance policy (1%).

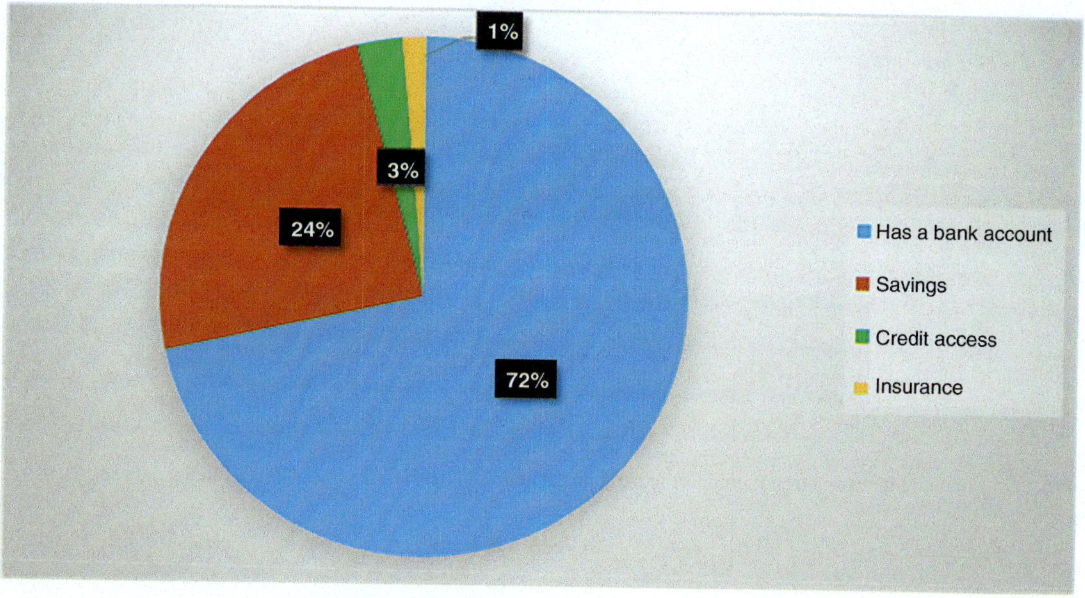

Fig. 8.1 Percentage distribution of main components of financial inclusion for Nigerian women. (Source: EFInA, 2020 Financial Inclusion Survey)

8.6 What Data Says About Opportunity to Drive Financial Inclusion for Women

Analysis of the data was explored across the four dimensions of financial inclusion identified (refer to Fig. 8.1) and over spatial features of urban and rural residence.

8.6.1 Sociodemography

The demography of women in Nigeria reveals less-inclusive indices when compared to men. Studies have shown that women are the major vulnerable groups being excluded from accessing financial services in many economies (Izquierdo and Tuesta 2015). The effects of socioeconomic backgrounds are seen to drive the other factors that drive financial inclusion (Adeleke 2021). This makes it important to consider sociodemographic characteristics as a baseline for creating the opportunity to access financial services. Sociodemographic factors such as age, sex, residence, educational status, and other variables are important as baseline requirements for adults to be involved in the formal financial system in any economy.

Being female has been a disadvantage for most Nigerians in accessing formal financial services. Data from the Global Financial Inclusion Survey (Global Findex), 2018, showed that women at the various dimensions of financial inclusion have had lower representation in the financial inclusion picture of the Nigerian system (see Fig. 8.2). Although access to financial services increased between 2011 and 2014 in Nigeria (30% to 44%), it reduced by about 4% between 2014 and 2017; the gap was higher for women. Access to financial services through a bank account was seen to have reduced from 34% to 37% between 2014 and 2017. The utilization component was even more dismal, with an already disadvantaged population sliding further below the national aggregate with a reduction in the percentage of women who saved by up to 7% between 2014 and 2017. In both cases, the data, when measured by access and utilization and compared with the overall average and male categories, suggests a form of low-level financial inclusion for women.

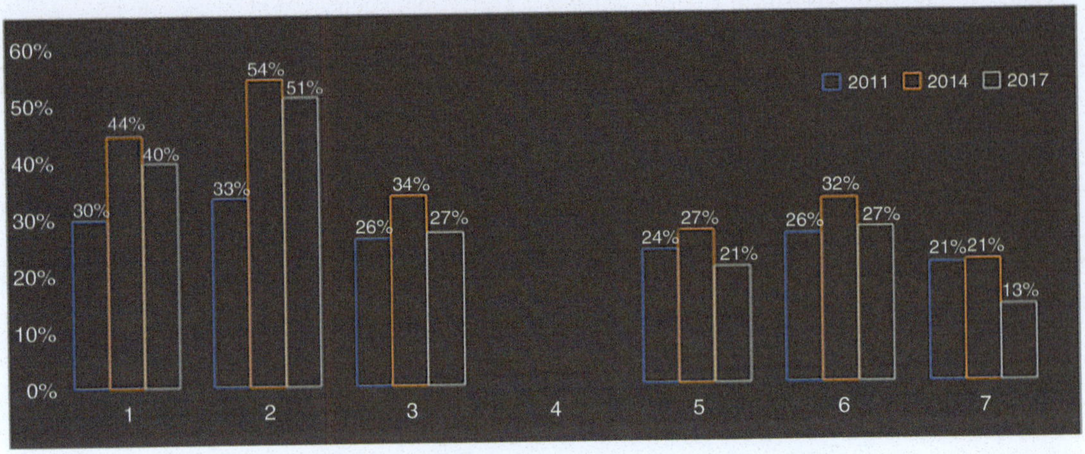

Fig. 8.2 Financial inclusion trend: comparing male and female in Nigeria. (Source: Global Findex data set, 2018)

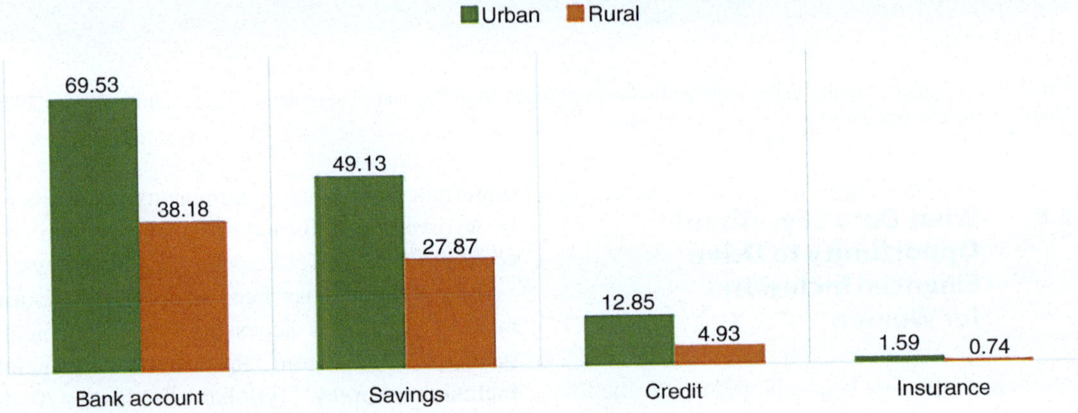

Fig. 8.3 Rural-urban differences in women's financial inclusion in Nigeria

Residence in rural areas is also a critical factor in providing opportunities for accessing financial services. The rural sector by default seems to be lacking in major support systems for financial services and thus translates to the level of financial inclusiveness of the residents. Differentials occur in access to financial services between and among urban and rural women in Nigeria. The EFInA data revealed that women in urban areas had better financial inclusion than rural women (See Fig. 8.3). The study, (Adewoyin et al. 2022) however, showed that certain critical variables such as education may enable rural women to have better access to financial services. The effect of sector of residence resonates in the remaining sections of this study, in a bid to understand differentials in the financial inclusion of women by residence.

The role of education is made obvious by the formality involved in financial systems. Data reveals (Table 8.1) that women with tertiary education were more likely to have a bank account (91.89%), savings in a formal institution (74.88%), an insurance policy (4.28%), and access to credit (19.49%). Variations across rural and urban residence revealed that rural women at all levels of education were less likely to be financially included than their urban counterpart. Education provides a baseline opportunity for women's capacity to access formal financial services (Sharif et al. 2022). Indeed, formal education is not the sole driver in this case; the lack of financial literacy is also seen as a barrier to gaining access to financial services in Nigeria (Atkinson and Messy 2013; Sabic-El-Rayess 2019; FINCA 2022).

Table 8.1 Education as an opportunity for financial inclusion among women in Nigeria

Education status	National				Urban				Rural			
	Bank account	Savings	Credit	Insurance	Bank account	Savings	Credit	Insurance	Bank account	Savings	Credit	Insurance
Non-formal	16.35	4.45	1.32	0.24	31.95	6.61	2.13	0.46	13.6	4	1.19	0.2
Primary	30.27	14.59	3.85	0.49	47.62	20.52	6.38	0.43	24.69	12.49	3.05	0.51
Secondary	57.3	34.62	7.87	0.68	72.65	44.49	14.48	1.02	49.42	29.19	4.69	0.5
Tertiary	91.89	74.88	19.49	4.28	95.02	77.52	18.64	4.26	88.18	71.61	20.35	4.31

Source: EFInA Survey, 2020

8.6.2 Sociocultural Context

The sociocultural context is a direct flow from sociodemographic characteristics of individuals that drive financial inclusion. Unfortunately, the culture of many developing countries appears to bring about a disconnect between preserving culture and real-life manifestation of development indices. Cultural barriers to women's autonomy and agency have been implicated in different forms of social, cultural, and physical discrimination (Demirgüç-Kunt et al. 2013). This calls for a review of the cultural landscape that allows women to make informed decisions about critical issues in their lives. In many ways, autonomy is correlated with the livelihood/economic situation of women; hence, women in labor force are more likely to be able to make decisions on how to spend their income, including in utilizing financial services (Ibtasam et al. 2018; Eckhoff et al. 2019). This bi-causal relationship between decision-making power and financial autonomy is important for sustaining the livelihoods of women and the welfare of their households (Hendriks 2019).

The process of developing livelihood and accessing financial services is intricately linked to the assets owned by individuals. In Nigeria, the cultural perception of women's property rights may have been a limiting factor in the opportunity stream for financial inclusion. Reports from the 2020 financial inclusion survey shows that only 54% of women had control over money in the household, while about 27% had the right to sell assets (EFInA 2021). These are indicative of low agency of women in terms of autonomy and freedom (Donald et al. 2020), which invariably translates to the extent of financial inclusion. The results in Fig. 8.4 show that the right to sell asset is translated to higher access and use of financial services across both urban and rural sectors in Nigeria. The spill-over effect of this in the livelihood structure of women shows how capital accumulation (savings and credit) is improved with women's right intact.

8.6.3 Livelihood Structure/Economy

This opportunity factor includes labor force participation, income inequality, poverty, and effect of the shadow economy. A stable economy has been stated as a precondition for a sustained financial inclusion. Generally, over 50% of Nigerians have been reported living below the poverty line, with the disaggregation showing the majority as female (National Bureau of Statistics 2019), across various cross indicators and with consequences for income inequality and financial inclusion accessing financial services. The effect of income in driving financial inclusion is, however, closely linked to how integrated the individual is in the labor force. On a national aggregate, Fig. 8.5 reveals a decreasing level of financial inclusion in general, but a lower baseline and a steeper change without labor individuals.

The attendant effects of low-income levels of women are a high level of income inequality as well as low agency for women. Without agency, women are cut off from accessing available financial activities or policies within their societies. Rural women are especially vulnerable in this placement. Rural women are usually found in the informal sector with weak and vulnerable capital bases. This weak base is a prerequisite for a fragile livelihood system in which most rural women are found.

Women in developing countries including Nigeria are increasingly involved in the formal economy; however, a large proportion obtain their livelihood within the informal economy (DiRienzo and Das 2021). The prevalence of women in the shadow economy is connected to the cultural and societal structures that have limited women's labor participation (Olu-Owolabi et al. 2020). The shadow economy is also a more attractive for rural women (Koolwal 2019)which may affect the extent to which they are able to access and utilize the financial market (Younas et al. 2022). Thus, creating opportunities for women to engage in the formal economy will also open up the opportunities to participate in the formal financial system in Nigeria (Ajide 2021).

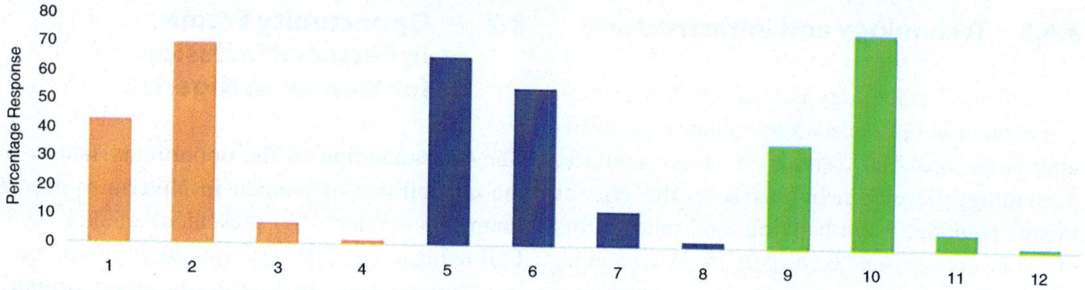

Fig. 8.4 Percentage distribution of financial inclusion by women with autonomy in Nigeria. (Source: EFInA Survey, 2020)

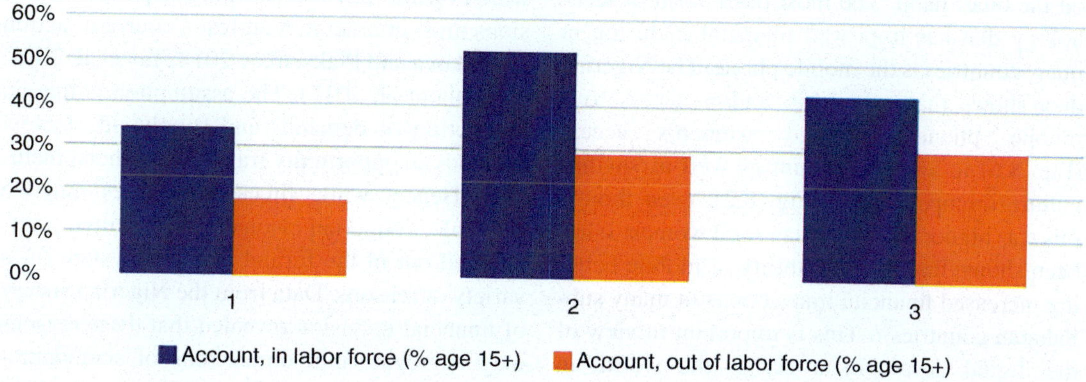

Fig. 8.5 Percentage distribution of account ownership by employment status in Nigeria. (Source: Global Findex Data, 2017)

8.6.4 Policies

Changing regimes brings changing macroeconomic and social policies in many countries. However, gender-responsive policymaking is still weak in developing economies. The system is such that consideration of women may be termed an after-thought of a policy development framework. This speaks to the need for gender-responsive budgeting at all levels of government, which is currently lacking in Nigeria's budgeting process. This lacuna effectively cuts out female-specific responses to policy formulation for women (Adeyeye et al. 2011). A gender-responsive budget involves gender resource allocation, gender-accessed budgets, and need-based gender budgeting, and which will effectively mainstream gender equality within the national budgeting cycles (Downes and Nicol 2020).

Certain financial policies have also been implicated in promoting financial exclusion. This could be by reducing the capability of financial systems to expand their service or by restricting effective demand for financial services on the part of consumers (Anarfo et al. 2020).. There may be a need to examine the effects of such banking policies as the bank verification number (BNV) in restricting access to banking services, in view of low-level socioeconomic status of women in Nigeria (Olaiya and Adekunle 2021). Flexible policies are, therefore, important to open the platform for an engagement with the financial system (Ezechukwu 2021). The Central Bank of Nigeria recognizes the need to be intentional in formulating and implementing financial inclusion policies with a gender perspective (Igwe and Adelusi 2021). However, the need for policy coherence and integration with a basic understanding of the response across gender groups may better provide a more effective baseline opportunity to gain access to and use financial systems.

8.6.5 Technology and Infrastructure

The effect of technology is clearly related to the infrastructural base needed to enhance the availability of financial service to the populace. Technology development has seen the rise of money banking, agent banking, and other forms of online transaction (Bayero 2015; Wayne et al. 2020). The platform of technology provides a means to host innovative financial instrumentation on the one hand and sustain the use of such on the other hand. The most basic form of technology that has improved financial inclusion in many countries is the mobile phone. The Nigerian data shows that ownership and/or access to a mobile phone increased women's access (Fig. 8.6) to a bank account as well as savings culture (compared with Fig. 8.2), even though this was higher for urban than rural women. It has been shown that the availability of mobile banking increased financial transactions in many sub-Saharan countries[69]. This is important in view of the digital development across the world, in which Nigeria has been on the low side of the digital divide (Adeleke 2021).

Thus, without the opportunity of a mobile phone, individuals are unable to take up this advantage, increasing the gender inequality in financial inclusion amidst other developmental issues (Treinen and Van Der Elstraeten 2017), creating a barrier to financial inclusion (LeFevre et al. 2020). In Nigeria, the ownership and use of mobile phones have been suggested as a means of empowering women and improving their income, hence participation in financial markets (Abubakar and Dasuki 2018). However, there is evidence of a digital divide between males and females, and even more importantly between urban and rural residents, such that the digital divide is wider between urban and rural residents, especially among women (Okunola et al. 2017). Bridging this divide will create a platform for which women are able to gain a first-level entry into the financial system in the country.

8.7 Opportunity Frame in Financial Inclusion for Women in Nigeria

The understanding of the opportunity matrix in the capabilities of women in Nigeria to access financial services is presented in Fig. 8.7. Following from data descriptions in the preceding sections, this study proposes an opportunity frame to drive financial inclusion for women in Nigeria. This follows an ecosystem model in which various physical, social, and policy dimensions must interact to manifest a course of action (Kabakova and Plaksenkov 2018; Tay et al. 2022; Okunola et al. 2017). The assumption is that an interaction of demand- and supply-side factors provides an opportunity frame for financial inclusion. Hence, while financial services may be available, Nigerian women have often self-selected out of the formal financial system for a variety of reasons. Data from the Nigerian survey of financial inclusion revealed that these reasons may aggregate into dimensions of sociodemographic, cultural, technological, economy, and policy factors. The interaction of these financial ecosystem dimensions procures a positive demand for financial service. These dimensions interact both directly and indirectly within this opportunity frame to give positive financial inclusion among women in Nigeria.

8.8 Policy Implications and Conclusion

8.8.1 Policy Implications

Creating opportunity for financial inclusion has been reflected as a matrix beyond gaining the access by increasing the number of financial services and/or making them affordable. There are supply-side and demand-side factors that create the agency for women to access financial services within the context of their needs. On the supply side, the need to create conducive environment

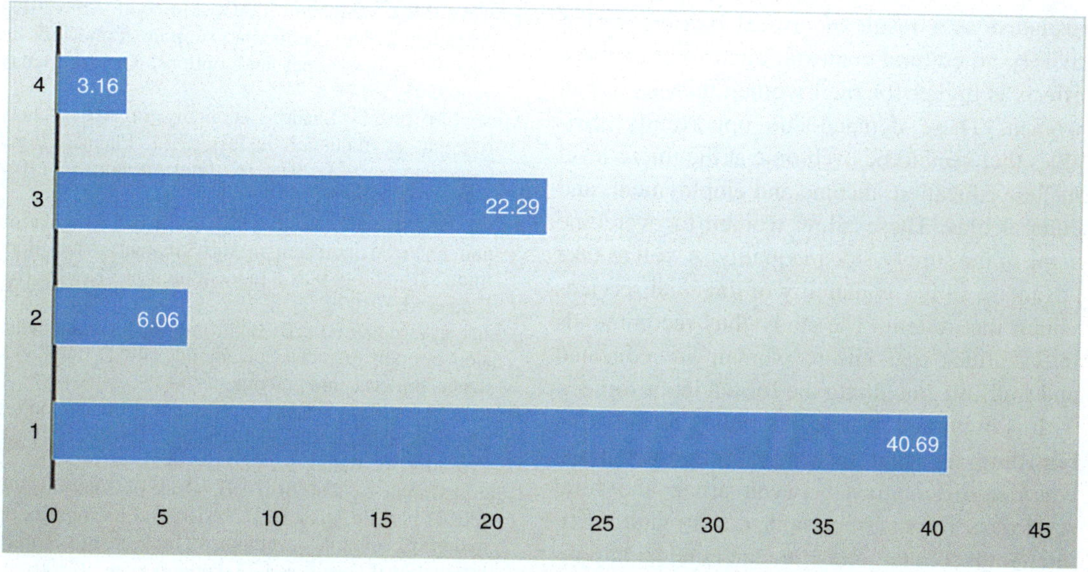

Fig. 8.6 Access to mobile phone and women's financial inclusion in Nigeria. (Source: EFInA Survey, 2020)

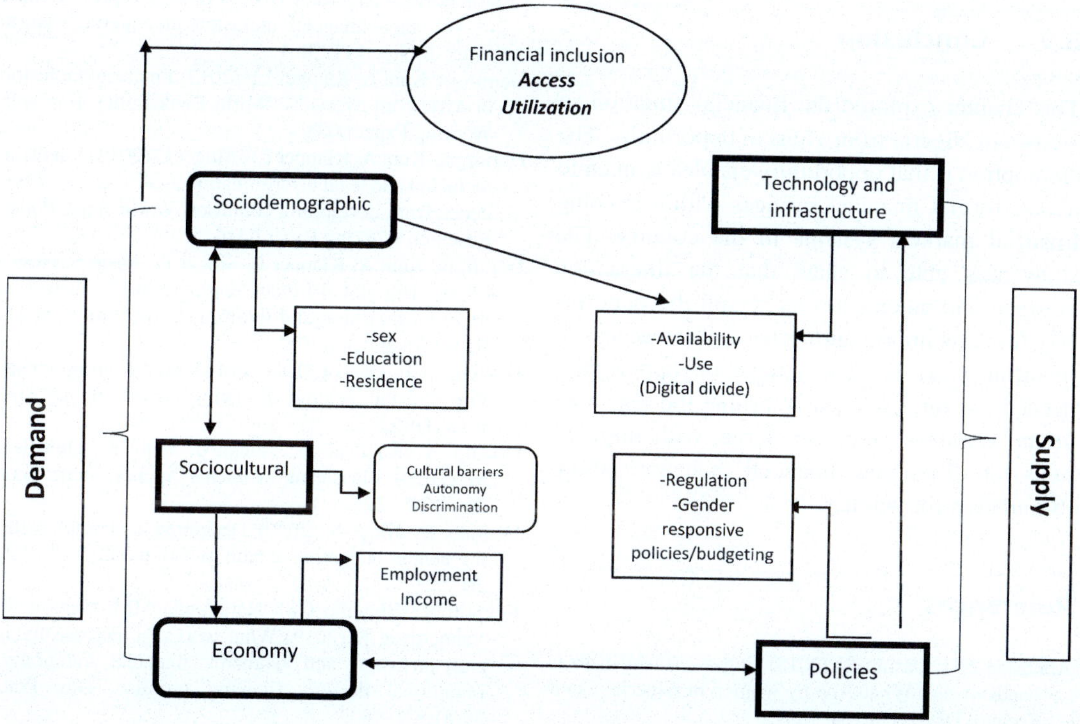

Fig. 8.7 Opportunity frame in financial inclusion for women in Nigeria. (Source: Author's conceptualization, 2022)

for policy and technology that are inclusive is critical. The study recommends a gender-responsive budgeting process which will ensure that the policy/institutional environment from which other opportunity factors are derived is made conducive for women.

Demand-side exclusion showed that sometimes women self-select into being financially excluded as a result of critical barriers within their socio cultural contexts. The extent of these effects is higher for rural women than for urban women. These demand-side opportunity variables that need to be overcome at the micro level include education, income and employment, and cultural bias. These allow women to overcome some of the supply-side inequality as well as take advantage of the availability of financial services within the system. The study thus recommends interventions that ensure women are educated and well-integrated into the formal labor force.

It is important to note that policy formulation regrading financial inclusion for women must consider differentials between urban and rural residences. This is important so as to gain maximum impact in targeting financial inclusion policies for women in Nigeria.

8.9 Conclusion

This chapter explored the financial situation of women in Nigeria from a lens of opportunity. The assumption is that opportunity enables women to access formal financial services within existing financial markets systems in the country. The study was able to show that the interaction between and among factors within the opportunity frame is important to give women access to financial services. The extent of opportunity needed for rural women to bridge the gap with urban women is, however, large, with implications for differential financial inclusion policy formulation for women.

References

Abubakar NH, Dasuki SI (2018) Empowerment in their hands: use of WhatsApp by women in Nigeria. Gend Technol Dev 22(2):164–183

Adeleke R (2021) Digital divide in Nigeria: the role of regional differentials. Afr J Sci Technol Innov Dev 13(3):333–346

Adewoyin Y, Nwosu IG, Ossai OG, Onuh J (2022) Rural-urban differentials in levels and correlates of financial inclusion among Nigerian women aged 18 to 49. J Popul Soc Stud [JPSS] 30:448–462

Adeyeye MM, Akinbami CAO, Momodu AS (2011) Adapting gender budgeting support framework in Nigeria: policy issues and options. Gender Behav 9(1):3595–3611

Ajide FM (2021) Shadow economy in Africa: how relevant is financial inclusion? J Financ Regul Compliance 29(3):297–316. https://doi.org/10.1108/JFRC-10-2020-0095

Anarfo EB, Abor JY, Osei KA (2020) Financial regulation and financial inclusion in sub-Saharan Africa: does financial stability play a moderating role? Res Int Bus Financ 51:101070

Atkinson A, Messy FA (2013) "Promoting financial inclusion through financial education:" OECD/INFE evidence, policies, and practice

Bayero MA (2015) Effects of cashless economy policy on financial inclusion in Nigeria: an exploratory study. Procedia Soc Behav Sci 172:49–56

Cabeza-García L, Del Brio EB, Oscanoa-Victorio ML (2019) Female financial inclusion and its impacts on inclusive economic development. In: Women's Studies International Forum, vol 77, Pergamon, p 102300

Central Bank of Nigeria (2018) National Financial Inclusion Strategy. Central Bank of Nigeria

Central Bank of Nigeria (CBN)/EFInA (2019) Assessment of Women's financial inclusion in Nigeria. Central Bank of Nigeria/EFInA

Demirgüç-Kunt A, Klapper LF (2012) Financial inclusion in Africa: an overview. World Bank Policy Research Working Paper 6088

Demirgüç-Kunt A, Klapper LF, Singer D (2013) Financial inclusion and legal discrimination against women: evidence from developing countries. World Bank Policy Research Working Paper 6416

Demirgüç-Kunt A, Klapper L, Singer D, Ansar S, Hess J (2018) The global Findex database 2017: measuring financial inclusion and the Fintech revolution. World Bank, Washington, DC

DiRienzo CE, Das J (2021) Formal female entrepreneurship and the shadow economy. J Econ Behav Stud 13(5 (J)):63–72

Donald A, Koolwal G, Annan J, Falb K, Goldstein M (2020) Measuring women's agency. Fem Econ 26(3):200–226

Downes R, Nicol S (2020) Designing and implementing gender budgeting–a path to action. OECD J Budg 20(2)

Eckhoff S, Kibombo GM, Natukunda MM, Pennotti C, Vandergaag K (2019) What works to increase financial inclusion and women's financial autonomy? Intentional designs showing promise. Dev Pract 29(8):974–987

EFInA (2021) EFInA access to financial services in Nigeria": 2020 Survey. June 3, 2021

Enhancing Financial Innovation and Access (EFInA) (2021) EFInA Access to Financial Services in Nigeria: 2020 Survey. June 3, 2021

Ezechukwu NV (2021) Regulating innovation for financial inclusion: lessons from Nigeria. J Afr Law 65(3):431–459

FINCA (2022) Financial literacy strengthens women's inclusion and empowerment. Retrieved online from https://finca.org/blogs/financial-literacy-strengthens-womens-inclusion-and-empowerment., 30/06/2022

Govindapuram S, Bhupatiraju S, Sirohi RA (2022) Determinants of women's financial inclusion: evidence from India. Ann Public Coop Econ 2022:1–28

Hendriks S (2019) The role of financial inclusion in driving women's economic empowerment. Dev Pract 29(8):1029–1038

Ibtasam S, Razaq L, Anwar HW, Mehmood H, Shah K, Webster J, Kumar N, Anderson R (2018) Knowledge, access, and decision-making: women's financial inclusion in Pakistan. In: Proceedings of the 1st ACM SIGCAS Conference on Computing and Sustainable Societies, pp 1–12

Igwe GU, Adelusi OA (2021) Role of financial regulators and supervisors in promoting Women's financial inclusion: evidence from the central Bank of Nigeria. CBN Bullion 45(1)

Izquierdo NC, Tuesta D (2015) Factors that matter for financial inclusion: evidence from Peru'. Aestimatio: the IEB International. J Financ 10:10–31

Kabakova O, Plaksenkov E (2018) Analysis of factors affecting financial inclusion: ecosystem view. J Bus Res 89:198–205

Kama U, Adigun M (2013) Financial inclusion in Nigeria: issues and challenges. Central Bank of Nigeria, Occasional paper 4:1–12

Kaur S, Kapuria C (2020) Determinants of financial inclusion in rural India: does gender matter? Int J Soc Econ

Koolwal GB (2019) Improving the measurement of rural women's employment: global momentum and survey research priorities. World Bank Policy Research Working Paper 8840

Kumar N (2013) Financial inclusion and its determinants: evidence from India. J Financ Econ Pol 5(1):4–19. https://doi.org/10.1108/17576381311317754

LeFevre AE, Shah N, Bashingwa JJH, George AS, Mohan D (2020) "Does women's mobile phone ownership matter for health?" evidence from 15 countries. BMJ Glob Health 5(5):e002524

National Bureau of Statistics. Poverty and Inequality in Nigeria (2019)

Okunola OM, Rowley J, Johnson F (2017) The multi-dimensional digital divide: perspectives from an e-government portal in Nigeria. Gov Inf Q 34(2):329–339

Olaiya AC, Adekunle OE (2021) Central Bank of Nigeria's Bank verification number policy and its effect on deposit money banks' customers; empirical evidence from Nigeria. Int J Bus Admin Stud 7(4):68–766. https://doi.org/10.20469/ijbas.7.10002-4

Olu-Owolabi FE, Amoo E, Samuel O, Oyeyemi A, Adejumo G (2020) Female-dominated informal labour sector and family (in) stability: the interface between reproduction and production. Cogent Arts Human 7(1):1788878

Ouma SA, Odongo TM, Were M (2017) Mobile financial services and financial inclusion: is it a boon for savings mobilization? ("[PDF] Mobile financial services and financial inclusion: is it a boon …"). Rev Dev Finance 7(1):29–35

Sabic-El-Rayess A (2019) Role of education in financial inclusion of poor and unbanked women in India. Indones J Learn Adv Educ (IJOLAE) 1(2):72–90. ("Amra Sabic-El-Rayess- Google Scholar")

Sarma M, Pais J (2011) Financial inclusion and development. J Int Dev 23(5):613–628

Sharif SP, Naghavi N, Waheed H, Ehigiamusoe KU (2022) The role of education in filling the gender gap in financial inclusion in low-income economies. Int J Emerg Mark

Tay L-Y, Tai H-T, Tan G-S (2022) Digital financial inclusion: a gateway to sustainable development. Heliyon

Treinen S, Van Der Elstraeten A (2017) Extending the benefits: gender-equitable, ICT-enabled agricultural development. The World Bank

United Nations (2022) Financial inclusion. Retrieved online from https://www.un.org/development/desa/socialperspectiveondevelopment/issues/financial-inclusion.html; 20th May, 2022

Wayne T, Soetan T, Bajepade G, Mogaji E (2020) Technologies for financial inclusion in Nigeria. Research Agenda Working Papers 2020(4):40–56

World Bank (2015) Overview: National Financial inclusion strategies. Retrieve from https://www.worldbank.org/en/topic/financialinclusion/brief/national-financial-inclusion-strategies. 20th May 2015

World Bank (2022) Universal Financial Access. Retrieved online from https://ufa.worldbank.org/en/ufa, 23/06/2022

Younas ZI, Qureshi A, Al-Faryan MAS (2022) Financial inclusion, the shadow economy and economic growth in developing economies. Struct Chang Econ Dyn

9 Financing Women and Gender Inclusion in Cote d'Ivoire's Peacebuilding Process

Omosefe Oyekanmi and Agunbiade O. Doyinsola

Abstract

Despite several national and international policies promoting gender inclusion in peacebuilding, there has been persistent under-investment in gender-inclusive peace in conflict and post-conflict settings with significant gaps in financing. The chapter examined the financing of women and their inclusion in the Ivorian peacebuilding process. It investigated women's role during the Ivorian civil wars and their contribution to building peace. Through the qualitative method, with the use of a case study research design, secondary data were sourced and content was analysed. The study revealed that from victims of sexual abuse to forceful conscription as combatants, the role of women during the civil wars remains complex. More so, through formal and informal structures, women were essential to the nation's efforts during the post-conflict reconstruction period.

The study, however, argues that with the disproportionate representation of women in peace processes, women lack opportunities to influence decision-making about financing priorities. Hence, the importance of involving women and financing them in the peacebuilding process cannot be overemphasized, given that women suffer more losses than men in conflict scenarios. Evidence gathered shows that women remain sidelined in peace processes and by extension were not adequately identified in most financial arrangements when compared to the men.

The study submits that though Côte d'Ivoire ranks as one of the fastest-growing economies in the world, its human development levels are low with some of the highest gender inequality rates in the world. This gender gap created by hierarchical structures of patriarchy inadvertently accounts for a low success rate of peacebuilding cases globally. The chapter argues that the non-inclusivity of women in the peacebuilding process affects decisions like inclusive financing which is significant to empowering women after a conflict. It is recommended, therefore, that the National Action Plans (NAPs) and other gender mainstreaming policies in peacebuilding should be reviewed and effectively implemented.

Keywords

Financing · Gender inclusion · Peacebuilding · Conflict

O. Oyekanmi (✉)
Pan African Women Studies Unit., Institute of Pan African Thought and Conversation, University of Johannesburg, Johannesburg, South Africa

A. O. Doyinsola
University of Ibadan, Ibadan, Oyo State, Nigeria

T. A. Ojo (ed.), *Women and Finance in Africa*, Sustainable Development Goals Series,
https://doi.org/10.1007/978-3-031-53337-2_9

9.1 Introduction

The poor financing of women and exclusion in peace processes has been one of the major barriers to successful peacebuilding. While women remain absent in peace processes and conflict resolutions, women's exclusion in peace negotiations and peacebuilding processes often translates to inadequate funding for women's early recovery needs. Considering that peace agreements involving women in their processes have a 35% likelihood to last for at least 15 years (UN 2022), increased women empowerment and funding in post-conflict scenarios has a greater propensity to improve the strength and longevity of peace agreements.

Following the two civil wars in Cote d'Ivoire and pockets of conflicts that sprung up thereafter, the impact on women is far-reaching. Aside from the loss of their husbands and the burden to cater to the family in the event of such loss, women have encountered human rights abuses and increased discrimination. Hence, the importance of involving women and financing them in the peacebuilding process, therefore, cannot be overemphasized, given that women suffer more losses than men (Plümper, and Neumayer, 2006). However, with women constituting about 49.1% of the Ivorian population, the exclusion of women in peacebuilding financing and sustainable solutions to peace is made worse. In 2004, the United Nations Operations in Cote d'Ivoire (UNOIC) launched residual functions (social cohesion, disarmament D–re-integration, security sector reform, etc.) to address the root causes of the conflict and build sustainable peace. At varying capacities, women functioned formally and informally in the peacebuilding process. Through popular support for peace negotiations, disarmament campaigns, community advocacy, and the creation of social, economic, and religious programmes that drive social cohesion, women frontally contributed to restoring peace in Ivory Coast. Again, during the conflict, as a way of humanitarian assistance, women created support systems for combatants and safe places for those fleeing the conflict. Nonetheless, women continue to be under-represented in decision-making spheres and marginalized in peace negotiation in Cote d'Ivoire (Traoré, 2019).

With the call for the full and effective implementation of the United Nations Security Council Resolution 1325 on Women, Peace, and Security, which accepted the inter-linkages across development, peace, security, and human rights, the principle was adopted in Côte d'Ivoire. The National Action Plan (NAP) for the implementation of Resolution 1325, in Côte d'Ivoire which is the first to be domesticated in Africa was for the period of 2008–2012 and later revised for the period of 2019–2023 (Côte d'Ivoire, 2023). Through progress meetings, the NAP does suggest that monitoring and evaluation will be carried out by committees, but it does not include a budget allotted for the stated initiatives.

Considering that the prerogative to implement the NAP is left to states, the 2015 Global Study on United Nations Security Council Resolution (UNSCR) 1325 affirmed that NAPs should include an allocated budget. However, most states failed to do so, which weakens accountability and impact (Peacewomen, 2022). The under-investment in the gender-inclusive peace process in conflict and post-conflict settings reveals the significant gaps in peacebuilding financing that make the implementation of women, peace, and security commitments more difficult. To this end, this chapter seeks to examine the effect of financing women and gender inclusion in the Ivorian peacebuilding process.

9.2 Methodology

The scope of this study is Cote d'Ivoire between 2004 when the UNOIC was established and 2017 when the UNOIC exited Cote d'Ivoire. The choice of Cote d'Ivoire is striking because Cote d'Ivoire is one of the most recent countries in Africa, where the UN peacebuilding was fully completed. Using a case study research design, through secondary sources of data collection, the chapter adopts a qualitative method, where the data collected was content analysed.

9.3 Women and International Peacebuilding

Globally and particularly in most African countries, women are marginalized and excluded from decision-making processes, principally in conflict-ridden situations of Africa where women are financially disempowered. In the aftermath of violent conflict and war, women unlike men in Africa have lesser protection or access to services, justice, economic security, and even citizenship. More so, delivery to meet basic needs and safeguard fundamental rights is void. Despite evidence proving the value of women's participation in peace processes globally, women largely remain excluded from conflict resolution and mediation.

In 2000, the United Nations Security Council (UNSC) unanimously adopted Resolution 1325 (The Resolution). This marked the first time the UNSC recognized the impact of armed conflicts on women and girls. The Resolution reaffirms the central role of women in conflict prevention, management, and resolution. The core element of the UNSC 1325 Agenda is the recognition of the importance of women's participation and representation in peace and security decision-making processes. As it stands, in fragile and conflict-affected countries, only 0.2% of bilateral aid goes to women's organizations (UN, 2021). More often than not, with the disproportionate representation of women in peace processes, women lack opportunities to influence decision-making about financing priorities. According to the United Nations (UN) Secretary-General in a 2009 report, funding women to boost early recovery from the impact of conflict is instrumental in addressing the longstanding gender inequalities. Hence, in comparison with men, women are largely marginalized and excluded in major areas of society, particularly in peacebuilding financing.

Between 1992 and 2019, on average, women accounted for 13% of negotiators, 6% of mediators, and 6% of signatories in major peace processes worldwide (Council on Foreign Relations, 2019), thus informing the wide financial gap between men and women in post-conflict recoveries globally. Again, the Security Council's resolution 2510 (2020), which endorsed the conclusions of the Berlin Conference on Libya (S/2020/63, annex I), called for the full, effective, and meaningful participation of women. However, the majority of Libyan representatives in the three intra-Libyan dialogue tracks on security, political, and economic issues that United Nations Mission in Liberia (UNSMIL) is facilitating are men (United Nations Security Council, 2020). Compared with 26% of other peace agreement types, only 11% of ceasefire agreements included provisions on gender equality, between 2015 and 2019 (Ibid). Besides indigenous women, women peacebuilders continue to face extreme abuse and violations during conflicts. The United Nations states that in 26 conflict-affected countries, between 2015 and 2019, there were 102 killings of women human rights defenders, journalists, and trade unionists (Norway in Geneva, 2020).

Bearing in mind that women are not adequately represented in peace negotiations and peacebuilding processes, women as vulnerable groups in society will continue to be exploited. In most cases, funds targeted at women's recovery programmes are either underfunded or diverted to other peacebuilding projects. A joint study conducted by UN Women and United Nations Population Fund (UNFPA) on some selected countries reveals the imbalance between funding and allocation of women's programme. According to the report, programmes tailored for women and girls were disproportionately underfunded despite receiving about 51% of the total funding requested to cover the needs of crisis-affected women and girls (UN Women, 2021) . However, where women inclusion in peacebuilding is encouraged, the peace sustained has a greater tendency to flourish. In the UN-cash-for-work women's initiatives in the Gaza strip, 250 women saw a 345% rise in their household incomes. Out of the 250 women, 71% noted a reduction in domestic conflict and 55% noted that their new benefit earned them access to household decision-making (UN Security Council, 2020).

Similarly, women are still excluded from disarmament and non-proliferation of weapons discourse, which is a pivotal function of peacebuilding. Though Disarmament, Demobilization and Reintegration (DDR) programmes take on a more male-centred approach, female combatants appear to be on the increase as observed in Eritrea, Mozambique, Zimbabwe Sierra Leone, Liberia, Uganda, and Rwanda. Aside serving as domestic staff and administrative or logistical personnel, women have assumed the role of armed combatants.

According to Hessmann-Dalaqua, Egeland and Hugo (2019), only one third of the participants at multilateral meetings on disarmament are likely to be women, with a lower probability of having women as heads of delegates. Again, since the existing relations between men and women in the society, particularly in the labour market, favour men, post-conflict policies that prioritize the privatization of basic service provisions negatively affect women (Klot, 2007). An assessment of the budget allocation for financing and addressing women's needs and issues states that of all the major indicators, the economic recovery and infrastructure sector was the least gender responsive in funding for peacebuilding. The report further states that only 7% of the total budget is allocated to activities or indicators that address women's needs and issues (Cueva and Kuonqui, 2012). However, peacebuilding processes now regularly acknowledge the participation of women, bearing in mind that beyond being victims, women also play a major role in violence. For instance, in Sierra Leone and Angola, the actual experience of women and girl fighters was not considered; they were only categorized as "dependents". This ultimately prevented them from receiving the benefits due to combatants (UN, 2021). The Global Acceleration Instrument (GAI) for Women, Peace and Security and Humanitarian Action instituted a new pooled funding mechanism, involving civil society at the level of governance and priority areas of funding (Peacewomen, 2022). Concisely, women are not only victims of conflict but may also play an active role in the violence; hence they must be involved in peace processes and negotiations like the men.

9.4 The Ivorian Conflict: Pre- and Post-conflict Period

From the 1960s, under Houphouët-Boigny's reign (the first President of Côte d'Ivoire), his philosophy was crafted in globalization, with attention captured on individualism over citizenship, what Akindès, (2003) refers to "Houphouëtian vision". Through his emphasis on individualism, he was able to promote pro-Western liberal policies and expand infrastructural development. Thus, the economic success and pragmatic course of the president sowed the seeds of the Ivorian miracle of the 1980s that attracted the influx of immigrants across West Africa into Côte d'Ivoire for greener pasture. Notably, without petroleum products, Côte d'Ivoire had one of the highest per capita incomes in sub-Saharan Africa, with an average real GDP growth rate of 7% between 1960 and 1980. However, this track record was not to be sustained, as the falling price of cash crops and the complexities waning the liberal land ownership laws between the allogenes (foreigners) and autochthones (indigenes) dwarfed existing harmony.

The drop in the demand for tropical products in the mid-1970s resulted in cutbacks of more than two thirds in the economic growth rate by the early 1980s. This is reflected in the spiking rise in unemployment, heavy debt burden, accrued in the lean years, and the offshoot of political challenges since independence. In 1989, the Houphouët-Boigny government was forced to accept dramatic measures such as the halving of cocoa (the mainstay of the Ivorian economy) and coffee prices paid to producers as well as the abandoning of the price guarantee system. As soon as economic growth dwindled, the skillful management of the country's socio-political balance became loose. With the price of cocoa decreasing, multiple cleavages arose to degrade the unquestioned paternal authoritarianism that had unified many tribes and groups for almost three decades. According to Akinde (2003), the leadership of the first president entailed a "model of compromise" which became porous at the event of his death in 1993. In effect, Houphouët-

Boigny's demise revealed the fragility in the politics and administration of Côte d'Ivoire. His death in 1993 cascaded the country into deeply divided ethnic and religious inequalities.

With elections in sight, nationalism and ethnic affiliation became a legitimate instrument for pushing politics. The political contest between the main political actors (Alassane Quattara of Northern extraction, Laurent Gbagbo, and Henry Konan Bedie of Southern origin) created a series of identity crises which on different accounts dichotomized the country into group inequalities. For instance, "Ivorite" a neologism was used as an identity clause for rallying political support among disenchanted southern citizens, who blamed foreigners for taking their jobs away and worsening their living conditions. These trajectories entrenched the divisions between the north and the south that birthed the two violent conflicts. According to the UN high commissioner for refugees, over 3000 Ivorian deaths were reported, with 250,000 refugees spread across Liberia, Togo, Guinea, and Ghana, with over one million internally displaced. These violent confrontations were informed by the disputed November 2010 elections, where the two major contestants, Alassane Quattara and Laurent Gbagbo claimed victory. Following the trail of events, the United Nations Operations in Côte d'Ivoire (UNOIC) was established.

The UNSC resolution 1528 came into force on 27 February 2004, birthing the United Nations Operation in Côte d'Ivoire (UNOCI) which was established to facilitate the implementation of the peace agreements and see to the end of the civil war (UN Security Council, 2004). With six residual functions, it was created as a full-fledged multi-dimensional peacekeeping force in 2004 (Novosseloff, 2018) and entered its second phase in 2010 after the disputed 2010 elections. The ascension of Quattara into power in 2011 heralded the third phase, and by 2013, the operation began to draw back and finally exited in 2017. Concisely, the Ivorian conflict is multi-dimensional, having economic and social influences, but mainly motivated by the political dimension.

9.5 Women's Role in the Ivorian Conflict and Peacebuilding Process

Women played several roles in the conflict itself and the peace processes that followed thereafter. They combined a blurred role of being victims and peacemakers. From victims of sexual abuse to forceful conscription as combatants, the role of women during civil wars remains complex. Numerous women and girls in Cote d'Ivoire suffered from sexual assault and were kidnapped to serve as combatants. While it is indisputable that there were women and girls among the warring factions on both sides, most of these women were kidnapped at gunpoint, subjected to sexual torture, and then forced to fight. The majority of the crimes against women in Ivory Coast were motivated by ethnicity, as government soldiers shot Burkinabe males and raped Burkinabe women and girls as well as residents in the country's north (Amnesty International, 2007). Women who were displaced or refugees in Côte d'Ivoire, notably those who had escaped the Liberian civil war in the 1990s, were frequently made more vulnerable to poverty, sexual assault, and exploitation. Both pro-government militias and armed opposition organizations, all employed mercenaries, including Liberians attacked women on a regular basis. Thus, women were subjected to sexual slavery and abuse, and in most instances must not be seen by the opposition groups, which could inform a sense of betrayal and death as punishment to the woman. It is worthy of note that while some women were forced to carry weapons, serve as partners for sex, and nurses for injured soldiers, other women participated voluntarily.

Beyond being victims of sexual abuse and molestation, women were also active as combatants. Just like the men, women participated at various levels as combatants. Most of them joined to uplift their identity, material needs, and security (Diallo, 2017). However, more of the female combatants functioned as auxiliary staff, and never held a command, but a few were in charge of arms stocking piling, and keeping.

Significantly, women's economic need was a major driver in their involvement as combatants. With the conflict came economic hardship, with women as the most hit. The conflict affected trading and subsistence income, and therefore informed loss of income. Rather than ideological, the economic motive, for instance, was one of the core reasons that cooks joined these groups, prompting them to move wherever work, living, and pay conditions were considered beneficial.

Hence, women's role in the Ivorian conflicts resonates around cooks, cleaners, smugglers of food, spies, hiding of weapons, fighters, and sexual gratification for men. While women's motive for joining groups is not different from the men, the roles women played were structurally defined, by societal roles. This largely shaped the outcome of women's gains after the wars were over. Not much changed for women regarding getting economic benefits, political appointments, and employment opportunities. The return to peace in Côte d'Ivoire meant a return to the sociocultural practices where patriarchy is a major determinant of financial inclusion and involvement in decision-making.

Bearing in mind women's involvement in virtually all areas of the conflict, women remain sidelined in the peace process and by extension were not adequately considered in any financial arrangements. During the 2003 Linas-Marcoussis Peace Accords and the 2007 Ouagadougou Political Agreement (OPA), there was no clear presence of women. At the Linas-Marcoussis Agreement in 2003, only one woman was present. Though the track II process of the Linas-Marcoussis Peace Accord shows evidence of women's group, their level of involvement remains unclear (Hudson, 2009).

Nonetheless, through civil society groups, international institutions, and donor agencies, some women at the level of individuals and community groupings mobilized themselves as conflict-prevention actors.

Following the peace works of Fanta Diamande, through the use of early warning and information networks, UN Women, UNICEF, and the Ministry of Women, Family, and Children (MFFE) launched a project in 2017 to empower women and girls as conflict prevention agents. By training in mediation, discussion, social cohesion, and the avoidance of gender-based violence, the programme attempted to lessen racial and political tensions at the local level. A network of more than 500 female peace mediators was established. Many of these women previously held prominent positions in their communities or belonged to grassroots civil society organizations. This platform was able to unite women from all ages and socioeconomic backgrounds in ten distinct areas of the nation (UN Women, 2021).

Women staged peaceful demonstrations demanding that the President pull out of the Paris Peace Meeting in 2003. Also, women made visits to French and American embassies in Abidjan to convey their objection to the Linas-Marcoussis Accord that was believed to have conceded too much to the rebels. Women collaborated with civil society groups and several informal channels to press on their demands in the peace process. Admittedly, women were actively excluded from formal peace negotiation processes when organizing peacebuilding initiatives. However, women were essential to the nation's efforts during the post-conflict reconstruction period, through informal structures.

9.6 Barriers Inhibiting Women's Involvement in the Peacebuilding Process and Financial Inclusion in Côte d'Ivoire

Women have been disproportionally represented in the financial and peacebuilding process, even though the impact of wars on them leaves devastating consequences. Despite women's concerted efforts to promote and sustain peace in their communities and spaces, certain factors hold sway over the continued limitation of women in formal arenas of the peace process. These factors spiral from pre-existing patriarchal structures, which perceive women as subjects to men. In virtually all spheres of society, sociocultural practices subject women to discrimination in Africa. Compared to men, women face stiff challenges in education, health, land access, financial empowerment, and political participation, which negatively impact their inclusion and acceptance as decision-makers in core areas of human development.

Inferring from Alani (Badmus, 2009), the masculine super-ordination/feminine subordination dichotomy is determined by sociocultural patriarchal traditions, which create barriers to female development in politics and decision-making that affect women in all sectors of life in West Africa, in other words, the sociocultural myths and beliefs that determine how people view men's and women's rights beginning in childhood are the foundation of patriarchy.

Though Côte d'Ivoire ranks as one of the fastest-growing economies in the world, its human development level is low, with some of the highest gender inequality rates in the world (Word Bank, 2017). In Côte d'Ivoire, women account for almost half of the population and have been sidelined in almost all areas of decision-making. The World Economic Forum's 2018 gender parity report places Côte d'Ivoire 131st out of 149 nations considered. This ranking is based on four factors: political engagement of men and women, educational possibilities, health, and opportunities for economic participation. Likewise, the 2019 United Nations Development Programme (UNDP) Human Development Index report, rated Côte d'Ivoire 165th out of 189 nations (UN Women, 2021). Social indicators show gender disparity in the educational sector at the secondary level where girls account for 42.7% as against 55.5% of boys. Again, for every 100,000 live births, there were 643 deaths in Côte d'Ivoire, upscaling maternal maternity (Ibid). Economically and socially, women rank below men in most indicators. The artificial creation of men's and women's roles that places the male gender as superior superstructure prevalent in African society is inadvertently mirrored in peacebuilding processes. Women are not formally and adequately involved in peacebuilding, regardless of national and internal policies promoting gender inclusion. For instance, in the DDR programmes, female combatants did not benefit from differential treatments (Roth, 2013).

Within the Ivorian workforce, women make up about 49.3% and yet are not adequately represented in decision-making. Most women do not have equal opportunities like the men in assessing formal jobs and are then forced to embrace the informal sector and agriculture. Thus, the employment rate of women stands at 9.4% (Christiaensen, and Premand, 2017). As such, economic empowerment for most women comes from the informal sector where women make up 62% of self-employed business owners. While women participate in a variety of economic sectors, such as in the retail, wholesale, apparel manufacture, domestic services, and hospitality industries, their participation is at a lesser percentage than men (ILO, 2012). This economic access, even though it presents an avenue for women to make half as much money as males do on average, still does not grant the right of entry into specific professions like the civil service and political office (CARE, 2019). Within the political space, women's participation in public office has been influenced by male domination. Politically, women have been marginalized majorly because most cultures in Côte d'Ivoire do not accept women as leaders in public spaces. As of February 2021, women accounted for only 11% of the members of the Ivorian parliament. Particularly, during the war and in violent settings, women face more precarious situations. In the event of losing a husband, land ownership supported by customary practice is bequeathed to the man's family without recourse to the woman. In situations, where funding may be available, the conditions attached to accessing such funds may prove extremely challenging. This is more difficult for women at the bottom spectrum of the society. In effect, these pre-existing variables within the societal realm dovetail with the structures that manifest in post-conflict peacebuilding management and essentially as it relates to financial inclusion. Based on a study by the World Bank Group and Global Findex Database between 2011 and 2017 on financial inclusion by gender in fragile and conflict-affected countries, more than one in three men own financial accounts compared to women, with only one in four women having a financial account (World

Bank, 2017). What this portends is that the financial inclusion of women in post-conflict peacebuilding programmes and services will become low, considering that owning a financial account is central to accessing loans and grants. According to the same data, women with access to financial accounts in Côte d'Ivoire for the year 2017 are less than 40% (Klugman, and Quek, 2018). Again, with this low number, opportunities for engagement with financial products and services after the conflict remain slim, as women without financial accounts are more likely to use those of family and friends or risk the chance of benefitting from such programmes.

Accordingly, with agriculture as the mainstay of the Ivorian economy and land as a core source of wealth, most women are disadvantaged through old sociocultural practice that assigns property only to the male gender. About 66% of the Ivorian land is used for agriculture with 43% of women working in the agricultural sector, yet going by the customary law, most women are debarred from owning landed properties. Hence, with land as a major channel for accessing credit, women are marginalized. Assuredly, women are limited in access to microcredit, and in the long run economic empowerment. The scenario that plays out during the peacebuilding process is no different. In most cases, post-conflict reconstruction targeting large-scale output is mainly dominated by men and benefits men more. However, support for women by the international community concentrates on microenterprises and microcredit. The effect of economic dis-empowerment on women's involvement in the peace process is brazing.

With the gender gap created by hierarchical structures of patriarchy, the inclusion of women in the peacebuilding process is still skeletal. This has invariably dampened the success rate of peacebuilding cases globally, considering that certain decisions like inclusive financing peculiar to empowering women after a conflict may not be incorporated into the peacebuilding framework, especially since evidence suggests that women's participation in peace negotiations improves the strength and longevity of peace agreements (Okia, 2020). The low ranking of the female gender at the social, economic, and political levels sprouts from the cultural delineation of the Ivorian society which places the male gender superior to the female gender. Conclusively, this age-long patriarchal belief has informed the limited access of women to finances and decision-making in peacebuilding process.

9.7 Conclusion/ Recommendation

Women are a major stakeholder in building peace and conflict resolution in war-torn states. Their involvement in conflict and peacebuilding stems from the angle of a victim to a combatant. In the event of conflict and war, they are sexually abused and molested. While women suffer so much loss during the conflict, they have been active in promoting and sustaining peace within their local community but majorly through informal channels.

Though national and international policies promoting the inclusion of women in the peacebuilding process abound, women active in the decision-making process are few and thus reflect the limited access to financial inclusion. The patriarchal system which places the male on a superior hierarchy to women is the major driver inhibiting the involvement of women in peacebuilding. Thus, the NAPs and other gender mainstreaming policies in peacebuilding in Côte d'Ivoire should be reviewed and effectively implemented. Deliberate and intentional policies influencing gender parity in the peacebuilding process must also be pursued. For instance, more funding should be stipulated and allocated to female peacebuilders and combatants. Without de-limiting the Ivorian culture, the Ministry of Social Cohesion in Côte d'Ivoire must re-orientate citizens on the equality of all persons as a basic component of human rights.

References

Akindès F (2003) Côte d'Ivoire: socio-political crises,'Ivoirité' and the course of history. Afr Sociol Rev/Revue Africaine de Sociologie 7(2):11–28

Amnesty International (2007, March 15) Cote d'Ivoire: targeting women: the forgotten victims of the conflict. Retrieved from https://www.amnesty.org/en/wp-content/uploads/2021/07/afr310012007en.pdf

Badmus IA (2009) Explaining women's roles in the West African tragic triplet: Sierra Leone, Liberia, and Cote d'Ivoire in comparative perspective. J Altern Perspect Soc Sci 1(3):808–839

CARE (2019) Contexte et Analyse de Politique, Action &Strategie de Plaidoyer pour le Programme «femmes enentreprise». Available at https://www.careevaluations.org/wp-content/uploads/CONTEXTE-ETANALYSE-DE-POLITIQUE-ACTIONSTRATEGIE-DE-PLAIDOYER-POUR-LE-PROGRAMME-%C2%ABFEMMES-EN-ENTREPRISE-%C2%BB.pdf

Christiaensen L, Premand P (2017) Côte d'Ivoire Jobs Diagnostic: employment, productivity, and inclusion for poverty reduction. Retrieved from https://openknowledge.worldbank.org/entities/publication/e0457bce-f510-514c-b075-00d88384cf1b

Côte d'Ivoire (2023) Retrieved from http://1325naps.peacewomen.org/index.php/cote-divoire/

Council on Foreign Relations (2019, January 30) Women's participation in peace processes. Retrieved from www.cfr.org/womens-participation-in-peace-processes

Cueva H, Kuonqui C (2012) What women want: planning and financing for gender-responsive peacebuilding. Retrieved from https://www.unwomen.org/sites/default/files/Headquarters/Attachments/Sections/Library/Publications/2012/10/WPSsourcebook-05A-PlanningFinancing-en.pdf

Diallo K (2017, December) When women take part in the rebellion: the ex-fighters from Ivory Coast. Noria Research. Retrieved from http://www.noria-research.com/women-fighters-ivorycoast/#:~:text=According%20to%20the%20Authority%20for,activities%20(cooking%2C%20cleaning%E2%80%A6

Hessmann-Dalaqua R, Egeland K, Hugo TG (2019) Still behind the curve: report prepared for the United Nations Institute for Disarmament Research. Geneva. Retrieved from https://unidir.org/files/publication/pdfs/still-behind-the-curve-en-770.pdf

Hudson H (2009) Peacebuilding through a gender lens and the challenges of implementation in Rwanda and Côte d'Ivoire. Secur Stud 18(2):312

ILO (2012) Enquête nationale sur la situation de l'emploi. Retrieved from https://www.ilo.org/surveyLib/index.php/catalog/7800/data-dictionary/FA_CIV_ENSE_2012_FULL/?offset

Klugman J, Quek Y (2018) Women's financial inclusion and economic opportunities in fragile and conflict-affected states: an overview of challenges and prospects. Georgetown Institute for Women, Peace and Security

Klot JF (2007) Women and peacebuilding: independent expert paper, commissioned by the United Nations Development Fund for Women and the Peacebuilding Support Office. United Nations, New York

Norway in Geneva (2020, October 30) Women, Peace & Security: UNSCR1325 is 20 years. Retrieved from https://www.norway.no/en/missions/wto-un/our-priorities/humanitarian-affairs/1325-20ar2/#:~:text=The%20UN%20has%20verified%20102,protective%20measures%20during%20the%20pandemic

Novosseloff A (2018) The many lives of a peacekeeping mission: the UN operation in Côte d'Ivoire. Retrieved from SSRN 3261285.

Okia A (2020) Strengthening women's role in building and sustaining peace. Retrieved from https://www.undp.org/speeches/strengthening-womens-role-building-and-sustaining-peace

Peacewomen (2022) Women, peace and security financing. Retrieved from http://www.peacewomen.org/WPS-Financing

Plümper T, Neumayer E (2006) The unequal burden of war: the effect of armed conflict on the gender gap in life expectancy. Int Organ 60(3):723–754

Roth N (2013) Gender, women, peace and security in West Africa' database. Retrieved from http://www.womencount4peace.org/sites/default/files/sites/default/files/publications/case%20study%20-%20CMI%20-%20%20Cote%20d%27%20ivoire%20%20Gender-Based%20Vilolence%20%20Conflict%20and%20The%20Political%20Transition%20-%20en%20-%2020130910.

Traoré SA (2019, September) Women at the table: peace and security in Côte d'Ivoire. Government of Canada. Retrieved from https://www.international.gc.ca/world-monde/stories-histoires/2019/cotedivoire-peace-paix.aspx?lang=eng

UN (2001, March) Gender perspectives on disarmament, demobilization and reintegration. Retrieved from https://www.oecd.org/social/gender-development/1896536.pdf

UN (2021, June) CSO survey on women, peace and security and humanitarian action (WPS-H A). Retrieved from https://wphfund.org/wp-content/uploads/2021/07/WPHF-CSO-Survey-on-WPSHA-June-2021_Final-2.pdf

UN Women (2021, March 10) This is what leadership looks like: Meet Fanta Diamande, a local peace mediator in Côte d'Ivoire. Retrieved from https://www.unwomen.org/en/news/stories/2021/3/what-leadership-looks-like-fanta-diamande

United Nations Security Council (2004) UN Security Council establishes peacekeeping operation in Côte d'Ivoire, unanimously adopting resolution 1528 (2004). Retrieved from https://www.un.org/press/en/2004/sc8012.doc.htm

United Nations Security Council (2020, September 25) Women and peace and security: report of the Secretary-General. S/2020/946. Retrieved from https://www.securitycouncilreport.org/atf/cf/%7B65BFCF9B-6D27-4E9C-8CD3-CF6E4FF96FF9%7D/s_2020_946.pdf

United Nations (2022) We need more women leaders to sustain peace and development. Retrieved December 20, 2022, from https://www.un.org/fr/desa/we-need-more-women-leaders-sustain-peace-anddevelopment#:~:text=The%20evidence%20is%20clear%3A%20wherever,to%20be%20unstable%20and%20unsustainable

World Bank (2017) Situation économique en Côte d'Ivoire, Et si l'Émergence était une Femme?. Retrieved from http://documents1.worldbank.org/curated/en/374581499668123584/pdf/117223-FREN

Digital Financial Inclusion for Women in Africa: Prospects and Challenges

10

Rifkatu Nghargbu and Fadila Jumare

Abstract

Digital finance is one of the fastest growing financial services in the world. The growth in digital finance can be attributed to an increase in the ownership of mobile phones which has triggered a tremendous increase in innovative digital tools, transactions and services worldwide. Digitization has contributed substantially to the GDP growth of many countries with the potential to further increase the GDP of all emerging markets by $3.7 trillion in 2025. This could increase new jobs, investments and overall productivity as e-commerce platforms are used to sell products and perform financial transactions across the world. Although digital finance is fast growing, many women in Africa do not have access to digital financial services, especially those living in poor and marginalized communities. They often do not have digital assets such as smartphones, internet services and also lack digital literacy to perform digital transactions. This chapter aims to discuss how women digital financial inclusion can be achieved in Africa, its prospects for women empowerment and likely challenges.

Keywords

Digital financial inclusion · Women · Information technology · Challenges and prospects

10.1 Introduction

The emergence of Information and Communications technology (ICT) and the increase in mobile smartphone usage within the last decade have brought about a surge of innovative digital tools and services across Africa with huge prospects for governments to build more inclusive economies. Apart from Africa, the G20 and the World Bank have also led the initiative to reduce poverty through digital financial inclusion in developing countries (Ozili 2018). Digital finance simply refers to financial services delivered through mobile phones, personal computers, the internet or cards linked to a reliable digital payment system. It encompasses all products, services, technology and/or infrastructure that enable individuals and companies to have access to payments, savings and credit facilities via the Internet (online) without the need to visit a bank branch or without dealing directly with the financial service provider (Ozili 2018). It is the impact of new technologies on the financial services industry (AFI 2016). Digital finance increases access to finance among poor individuals and

R. Nghargbu (✉) · F. Jumare
National Institute for Legislative and Democratic Studies, National Assembly, Abuja, Nigeria

T. A. Ojo (ed.), *Women and Finance in Africa*, Sustainable Development Goals Series,
https://doi.org/10.1007/978-3-031-53337-2_10

reduces the cost of financial intermediation for banks and financial providers. It also serves as the channel the government uses in increasing aggregate expenditure. Due to digital technology, the traditional way of providing banking services is being transformed to using a variety of products, applications and business models which makes banking transactions easier and more accessible. In addition, mobile applications, social networks, distributed ledger technology and cloud computing introduced new ways of banking and investment as well as new services and business models which people adopt to improve their finances (European Commission 2023).

Digital financial inclusion involves using digital means to financially reach out to the people who are financially unreached and underserved with digital financial services that meet their needs at a cost that is more sustainable to both customers and providers (The World Bank 2022). Digital financial inclusion can also be defined as a sustainable provision of affordable digital financial services that bring the poor into the formal economy (The World Bank 2022). This involves an increase in the number of the poor that have access to formal financial services through formal bank accounts with the aim of empowering them, which contributes to poverty reduction and economic growth (Ojo 2022). These individuals have greater opportunity to initiate businesses and carry out transactions that will enhance their income through digital means at a convenient and very low-cost rate. Digital financial platforms also serve as a means to increase savings and investment, services, credit, insurance and investment products which propel growth (Lyman and Kate 2015).

Digital financial inclusion is becoming an important phenomenon globally because of its prospect of reducing poverty, especially among women. Globally, women perform two thirds of labour but earn far less income; they also make up the most vulnerable and poorest worldwide (OXFAM International 2022). In Africa, women are the most economically active and represent half of the population (BSR 2016). They are also known to be one of the main pillars of economic development in Africa. Many women in Africa are financially excluded, with a wide gender gap in terms of income and social status. Despite the gender policy initiated in 2005 to facilitate the implementation of national and international conventions and laws in support of gender equality, wide gender gaps still exist. Many women in Africa are not open to opportunities for professional, technical and managerial jobs like their male counterparts. Women in Africa typically occupy the worst-paid jobs and face discrimination in accessing better opportunities that will improve their economic, political and social status. Most women in Africa work as petty traders and small-scale farmers, which are largely for subsistence. However, the advent of digital finance has prospects in providing the African government and policymakers the opportunity to empower women through digital financial inclusion. Digital finance can help to bridge the gender gap in income. Bridging gender gap is one of the major targets of the SDGs. Data from UN Women and UN Department of Economic and Social Affairs Gender Snapshot 2021, on the progress of the Sustainable Development Goals (SDGs) shows that by 2030, digital financial inclusion initiatives are capable of lifting over 150 million women and girls from poverty (Alliance for Financial Inclusion 2021). The objective of this chapter is to discuss the prospects and challenges of using digital financial inclusion to boost women income in Africa. Following this introductory section, the chapter contains seven sections; Sect. 10.2 discusses the trends in digital payments in Africa, Sect. 10.3 examines the issue of supporting infrastructure for digital finance, while Sects. 10.4, 10.5, 10.6 and 10.7 discuss the history of financial flows in Africa, entrepreneurship and digital finance in Africa and prospects and challenges for digital financial inclusion for women empowerment in Africa as well as conclusions and recommendations.

10.2 Trends in Digital Payments in Africa

The revolution in digital finance in Africa started with the introduction of mobile banking by telecom operators in 2007 known as "mobile money" (Safae 2020). In Kenya, mobile money such as M-Pesa improved the economy of Kenya such that poverty in Kenya was reduced by 2%, taking about 194,000 Kenyan households above the poverty line, with female-headed households being the greatest beneficiaries (McKinsey and Company 2022). Over the past decade, ownership of accounts at a bank or regulated institutions such as microfinance and mobile money services increased immensely. As a result of mobile money, 33% of adults in sub-Saharan Africa now have a mobile money account between 2014 and 2021 (Hare 2022). Africa's mobile money in terms of prospects is expected to grow by 20% per year, reaching around $40 billion by 2025 (McKinsey and Company 2022). This growth by 2025 may not be uniformly achieved across the continent due to disparity in infrastructure readiness and e-commerce penetration. For instance, Egypt, Ghana, Kenya, Nigeria, and South Africa have experienced faster growth in digital finance compared to other countries in Africa due to the fact that they are developing the appropriate infrastructure and relevant policy frameworks for e-commerce. Women can key into the growth in digital finance to improve their savings and payments methods for the overall improvement of their welfare.

10.3 Supporting Infrastructure for Digital Finance

Many African countries have focused on infrastructure such as procuring 3G and 4G networks and prioritizing other digital infrastructure in their national policies. There are three key infrastructures that enhance the growth of digital finance which African countries should focus on: biometrics, Internet connectivity and access to electricity. Many African countries in recent times adopted biometrics technology to enhance identity capture by measuring people's unique physical features, such as fingerprint and facial recognition for the purpose of auditing, voting and impact assessment. Biometrics is also used to dictate and curtail fraud, improve data security and as serve as a source for documentation/identification of citizens. Although Africa is making progress in adopting biometrics, there are several barriers to greater adoption of biometric technologies in the continent. Data privacy and level of protection of sensitive personal data is a major challenge as only 24 countries in the continent have introduced laws and regulations to protect personal data (Toesland Finbarr 2021).

In terms of Internet connectivity, the Internet in Africa is highly limited due to a lower penetration rate when compared to other parts of the world. The World Bank estimates that Africa will require an investment totalling $100 billion for every citizen to have Internet by 2030 (Mwakideu 2021). Only 40% and 0.4% of persons in Africa have access to the Internet and fixed-broadband subscription, respectively (Adepetun 2022). Majority of Internet users access Internet through mobile broadband. For instance, in 2020, Kenya had an Internet penetration of approximately 85.2% due to the introduction of M-Pesa, which is a mobile wallet provider (Adepetun 2022). Nigeria also has one of the largest mobile Internet users in Africa. Over 74% of web traffic is generated via smartphones and only 24% via PC devices (Wikipedia 2023). Obstacles to the accessibility of Internet services in Africa include low levels of computer literacy, poor infrastructures and high costs of Internet services as well as high cost of data for most people (Wikipedia 2023).

Access to electricity is also very low in Africa compared to other parts of the world, especially in rural areas. In about 42% of the total population in Africa, only 8% of rural residents (Puliti 2022) have access to electricity (Puliti 2022). Lack of access to regular and sustainable electricity is also a major setback, especially in rural areas that are not connected to power grids and frequent blackouts in major urban areas. Insufficient electricity in Africa has the potential to reduce the growth of digital finance. Women are not faring better in terms of biometrics,

Internet connectivity and electricity in Africa. There is need for governments in Africa to invest extensively to support the growth of these important infrastructures if Africa experiences growth in digital finance.

10.4 History of Digital Financial Flows in Africa

Innovations in digital financial services sector in Africa can be traced over a decade ago as the introduction of Internet banking, multi-currency cards and mobile money facilitated many financial transactions. Many cashless payments and remittances such as all forms of bill payments, purchasing airtime, receiving and sending money via Internet transfer, storing/depositing money through the e-wallet and purchasing other valuables have increased substantially within the last decade. In Kenya, 31% of the country's GDP is transferred through phones (Mims 2013), and M-Pesa in Kenya is one of the most utilized methods for workers in the city sending remittances to relatives in rural areas. Nigerian and South African economies are also good examples where cashless payments and remittances are growing. In South Africa, around 30–35% of retail payments were electronic and 60–65% were cash payments (Matambo and Schaefer 2013). In Africa, mobile money created an increase in remittances and displaced less formal channels of sending and receiving money. The African continent accounts for 70% of the world's mobile money value (Onyango 2022). About 53% of adults sent or received domestic remittances in 2021 (Matambo and Schaefer 2013) as mobile technology enables the transfer of money more cheaply and quickly to locations where it would be impossible without the technology.

Women in Africa are, however, limited in terms of digital financial transactions compared to men. This is because men utilize mobile money more regularly than women; they also tend to transfer money and buy airtime more often than women. Many factors have continued to hinder women from enjoying equal access to digital finance the same way as men. These include limited access to phones and Internet and lack of profitable employment and digital skills (Chamboko et al. 2018). In terms of access to smartphones and Internet, men are more likely to own a smartphone than women, especially in rural areas, and when women do own a mobile smartphone, their phones are usually older and cheaper than those used by men. In this regard, women's access to and use of mobile phones is very important if women will be digitally and financially included. The improvement in women ownership of phones should be focused on smartphones that will encourage the use of digital technology to enhance digital financial inclusion. Also, women should be encouraged to participate in training where they will be thought on how to use these phones for digital financial transactions (Chamboko et al. 2018).

10.5 Entrepreneurship, Digital Finance and Women in Africa

The growth of digital tools and e-payments has continued to increase the growth in online businesses, thereby fostering the growth of entrepreneurship. According to the Global Entrepreneurship Monitor, sub-Saharan Africa has the highest percentage of women entrepreneurs in the world, with 26% starting and managing a business on the continent (UNESCO 2021). Ghana is said to be in the lead for more female entrepreneurs than any other country, with 46% of businesses being owned by women (UNESCO 2021). In Africa, it is about 3.2% which is twice the global average of 1.6% (World Economic Forum 2022). However, women are underrepresented and marginalized in digital technology entrepreneurship due to the low global participation of women in technology sectors when compared to men. One of the reasons for the low participation of women in Africa is mobile Internet access. About two thirds of women in sub-Saharan Africa do not use the Internet, with gender gaps of about 41% and 36%, respectively. This is more prominent in rural areas and among women with less education (Omoruyi 2023). In Africa, the biggest challenge to mobile Internet access is owning a smartphone (GSMA 2019). Women often have less access to Internet and mobile smartphones due to affordability and low literacy level, especially in rural areas. Many women and girls do not have the opportunity to access training and education in science, technology, engineering and mathematics (STEM) that promotes general digital literacy. As such, fewer women have less critical digital skills needed to participate and compete digitally in the labour market. This dampens job growth and entrepreneurship opportunities for women. Women in rural areas and from poor families experience these barriers more than women in urban areas (Porfido and Marks 2020); in addition, female entrepreneurs also have challenges accessing finance to develop digital businesses.

To reduce these challenges and encourage women digital entrepreneurship, there are several ways women can be assisted; these include the promotion of STEM and literacy for girls. African governments can prioritize STEM education and digital literacy to empower women with transferable skills needed to thrive. Women can also be supported with affordable and secure broadband Internet access and web-based online platforms to access digital entrepreneurship skills. In addition, making data bundle affordable for women will improve Internet service quality and availability. Women can be encouraged to acquire digital skills to effectively use technology and digital tools, take advantage of the Internet and harness legitimate online opportunities. African governments can promote digital skill acquisition for girls and women by prioritizing education in ICT subjects. Women can also be empowered to be developers of digital tech and creators of digital tools and online content (Iraoya and Gbadebo 2022).

10.6 Prospects for Digital Financial Inclusion for Women Empowerment in Africa

The advent of digital financial services in recent times has increased the opportunities for millions of women across the globe such that more than 240 million women now have accounts with a financial institution or mobile money service (Global Partnership for Financial Inclusion 2020) compared to about 10 years ago. Several online trading and investment platforms provide opportunities for women to improve their income. This has enhanced women's ability to withstand financial, economic and health shocks (Global Partnership for Financial Inclusion 2020). The emergence of digital finance in Africa has impacted women in Africa positively (Ojo 2022) with several milestones and successes in the continent. Financial inclusion especially in mobile money adoption increased from 23% to 43% between 2011 and 2017 with East Africa taking the lead, followed by West and Central Africa. During and after the pandemic in 2020, the continent also had one of the fastest growth rates in mobile money with about 55–60% of the adult population which is expected to reach 634 million by 2025 (Jaiyeola 2022). Bank accounts rose in Senegal by 600% with higher growth in countries like Kenya where women's financial inclusion rose by about 80% with mobile money increasing women-headed homes savings substantially (Alliance for Financial Inclusion 2021). Digital financial services allow for easy, fast and flexible financial transactions across multiple channels. Ghana is one of the countries that has made use of digital finance judiciously for women empowerment. Ghana has one of the top three positions in the 2020 MasterCard Index of Women Entrepreneurs (MIWE) with the most women business owners (Alliance for Financial Inclusion 2021). The Bank of Ghana increased transaction limits on digital financial services and facilitated digital savings, investments, bill payments and online merchant payments (Alliance for Financial Inclusion 2021). These measures enabled customers to restructure their loan repayments and also provided credit to those in need. As of 2020, there were 17.5 million active mobile accounts in Ghana, representing an 84% growth from 2019; the largest growth beneficiaries were women (Alliance for Financial Inclusion 2021). In Kenya, the Central Bank of Kenya removed online transaction charges as well as online transaction limits to further help vulnerable households mainly women during the pandemic. The central bank also reduced the turnover tax rate from 3% to 1% for all MSMEs, with women owning half of the MSMEs. These efforts by the central bank increased the number of digital transactions, from 1.7 billion from January to November 2019 and then increased to 4.4 billion over the same period in 2020 (Alliance for Financial Inclusion 2021). This increase reflects a growth of 160.45%, with women benefiting significantly from digitalization in the country (Alliance for Financial Inclusion 2021).

In Nigeria, the 2020 MasterCard Index of Women Entrepreneurs (MIWE) identified Nigeria as one of the African countries with large growth in digital finance and e-commerce with a great deal of women participation. In enhancing entrepreneurship in Nigeria, the central bank took advantage of the growth of mobile smartphone utilization and e-payment to advance the course of entrepreneurship in Nigeria. The central bank is also currently working on a National Fintech Strategy in collaboration with Alliance for Financial Inclusion (AFI), to consolidate on digital economy in Nigeria (Alliance for Financial Inclusion 2021). The central bank provided guidelines for mobile money services. This in itself is a practical strategy for financial inclusion which is the driving force for closing the financial inclusion gender gap in Nigeria.

In Africa, digital finance has several advantages which have made it easier to serve as a tool for financial inclusion for governments and policymakers. These include easy access to finance for the poor at lower cost of financial intermediation for banks and fintech providers and increase in aggregate expenditure for governments. Digital platforms enable women to conveniently engage economically from cash-based transactions to formal digital financial transactions on secured digital platforms. This saves time and provides opportunity for women to engage in economic activities while taking care of their families at the same time. Digital finance provides convenient access to a diverse range of financial products and services as well as credit facilities which increases aggregate expenditure and national income.

10.7 Challenges

Despite the numerous prospects in digital finance, there are still limitations in women accessing the benefits. The limitations range from barriers to access to identification documents and lack of mobile smartphones, computers, tablets and digital skills. Access to smartphones is increasingly important with the rise of apps and e-commerce. Increasing women's access to smartphones and Internet is also critical to grow more inclusive digital economies. Reliable electricity is also a challenge, especially in rural areas.

Many women in Africa lack access to finance. African women are subjected to laws and norms that undermine women's right to participate in the labour force, control assets and establish and access funding to grow formal businesses. This contributes to limitations to digital financial inclusion. Digital mobile money has limited interoperability across digital financial service providers and an inability to prove one's identity (Gandh 2019). A lack of interoperability across service providers limits interpersonal customer transactions, leading to underutilization of these services (Alliance for Financial Inclusion 2021). Additional problems such as high conversion fees and having to travel long distances to convert digital money to cash limit the use of mobile money (Alliance for Financial Inclusion 2021). Many women participate in low-paying economic activities, which reduce their financial capacity to engage in digital financial transactions. In addition, competing demands on women's time related to unpaid domestic work are a serious constraint as women have little or no time to engage in other productive ventures. Another challenge is the lower rates of mobile phone ownership and lower rates of digital literacy among women which limit women's chances of being financially included. Internet infrastructure is also a major challenge in Africa since most people in Africa use mobile Internet for online financial transactions. African countries have the lowest mobile connectivity index scores in the world (Bill and Melinda Gates Foundation 2019a, b). This affects mostly women, the poor and many people in rural areas. Another challenge is dominance over women's financial activities by their partners. This limits their investment potential, which reduces their financial independence. Social norms may also be a contributory factor that influences women's financial inclusion. Cybercrime and Internet fraud is another factor that can discourage women to engage in digital financial transactions. Another serious challenge that has affected women's financial inclusion is the limited focus that government policy has on women's financial inclusion. Financial inclusion strategies and policies fail to consider women's perspectives and needs. In many cases, government programmes focus on specific empowerment programmes and population rather than a holistic approach that is targeted at solving women's problems. A more comprehensive view of policy across geographies and socio-economic dimensions will significantly support women financial inclusion and close the gender gap. To achieve this, sex-disaggregated data to inform policy and track progress towards financial inclusion goals is important. There must be a deliberate effort by financial service providers to collect sex-aggregated data on their clients which is currently not provided.

10.8 Summary, Conclusions and Recommendations

The chapter discussed the prospects and challenges of digital financial inclusion for women in Africa. The chapter shows that in Africa, digital finance has several advantages that are being harnessed, especially by government and policymakers. Digital platforms enable women to conveniently engage economically from cash-based transactions to formal digital financial transactions on secured digital platforms. This saves time and provides opportunity for women to engage in economic activities while taking care of their families. Digital finance provides convenient access to a diverse range of financial products and services as well as credit facilities which increases aggregate expenditure and national income. However, there are challenges associated with it; these include barriers in access to identification documents, lack of mobile phones and digital skills, lack of access to finance, lack of Internet infrastructure, cybercrime and Internet fraud and limited focus that government policy has on women's financial inclusion strategies which fail to consider women's perspectives and needs, lack of access to identification documents and limited digital financial infrastructure (Alliance for Financial Inclusion 2021).

To fully take advantage of digital finance in Africa, governments and policymakers must strategize to ensure that digital financial services support sustainable development objectives of achieving gender equality and empowerment of women and girls. This can be achieved by ensuring that digital financial services and platforms are affordable, easy to use, accessible and transparent. Access to mobile money accounts and support to identity system and documents will also help women to be more financially inclusive. Increasing women's access to smartphones and Internet is important in ensuring women inclusiveness in digital finance. Private sector support is also needed in terms of facilitating smartphone ownership.

References

Adepetun A (2022) Europe leads as only 40% of Africa has Internet access https://guardian.ng/technology/telecoms/europe-leads-as-only-40-of-africa-has-internet-access/

AFI (2016) Digital financial services basic terminology including acronyms for Dfs transactions. Guideline note developed by the Afi Digital Financial Services (Dfs). Guideline Note No. https://www.afi-global.org/wp-content/uploads/publications/2016-08/Guideline%20Note-19%20DFS-Terminology.pdf

Alliance for Financial Inclusion (2021) Africa's Financial Regulators Use Digitalization to Reach Women. Available at https://www.afiglobal.org/newsroom/blogs/africas-financial-regulators-use-digitalization-to-reach-women/

Bill and Melinda Gates Foundation (2019a) Women's digital financial inclusion in Africa. Available at https://www.greenfinanceplatform.org/research/womens-digital-financial-inclusion-africa JULY 2019

Bill and Melinda Gates Foundation (2019b) A G7 partnership for women's digital financial inclusion in Africa. Report prepared at the request of the G7 French presidency. Available at https://docs.gatesfoundation.org/Documents/WomensDigitalFinancialInclusioninAfrica_English.pdf

BSR (2016) From women's economic participation to empowerment: insights from Ghana. Report from BSR (2016). Online material derived from https://www.bsr.org/en/our-insights/blog-view/from-womens-economic-participation-to-empowerment-insights-from-ghana

Chamboko R, Heitmann S, Van Der Westhuizen M (2018) Women and digital financial services in sub-Saharan Africa: understanding the challenges and harnessing the opportunities. World Bank Report. Available at https://documents.worldbank.org/pt/publication/documents-reports/documentdetail/139191548845091252/women-and-digital-financial-services-in-sub-saharan-africa-understanding-the-challenges-and-harnessing-the-opportunities

European Commission (2023) What is digital finance? Available at https://finance.ec.europa.eu/digital-finance/what-digital-finance_en

Gandh D (2019) Increasing financial inclusion for women in Africa. Available at https://www.brookings.edu/blog/africainfocus/2019/07/24/figure-of-the-week-increasing-financial-inclusion-for-women-in-africa/

Global Partnership for Financial Inclusion (2020) Advancing women's digital financial inclusion. Report prepared for the G20 Global Partnership for Financial Inclusion by the Better Than Cash Alliance, Women's World Banking, and the World Bank Group. https://www.gpfi.org/sites/gpfi/files/sites/default/files/saudig20_women.pdf

GSMA (2019) Women, connected: the mobile gender gap report 2019. GSMA, London. Available at from https://www.gsmaintelligence.com/research

Hare A. (2022) Digital payments in sub-Saharan Africa: trends over the past decade. Available at https://www.povertyactionlab.org/blog/10-3-22/digital-payments-sub-saharan-africa-trends-over-past-decade-0

Iraoya O, Gbadebo O (2022) Digital financing frameworks, gendered entrepreneurship and the new equation. In: African economic conference, Balaciava, Mauritius 9–11. https://aec.afdb.org/en/papers/digital-financing-frameworks-gendered-entrepreneurship-and-new-equation-478

Jaiyeola T (2022) Nigeria's population favourable to mobile money industry growth. Punch Newspaper. https://punchng.com/nigerias-population-favourable-to-mobile-money-industry-growth-kanyeki/

Lyman T, Kate L (2015) What is digital financial inclusion and why does it matter. CGAP Blog. https://www.cgap.org/blog/what-digital-financial-inclusion-and-why-does-it-matter

Matambo G, Schaefer S (2013) Toward the cashless economy in Africa. Africa Frontier Forum. Available at https://newsroom.mastercard.com/wp-content/uploads/2014/02/Africa-Frontiers-Forum-How-Technology-is-Changing-Financial-Services-i....pdf

McKinsey and Company (2022) The future of payments in Africa. Available at https://www.mckinsey.com/industries/financial-services/our-insights/the-future-of-payments-in-africa

Mims C (2013) 31% of Kenya's GDP is spent through mobile phones. Available at https://qz.com/57504/31-of-kenyas-gdp-is-spent-through-mobile-phones

Mwakideu C (2021) Can Africa reach universal internet access? Digital World Africa https://www.dw.com/en/can-africa-achieve-universal-internet-access-by--2030/a-59729090

Ojo TA (2022) Digital financial inclusion for women in the fourth industrial revolution: a key towards achieving sustainable development goal 5. Afr Rev 14(1):98–123

Onyango S (2022) Africa accounts for 70% of the world's $1 trillion mobile money market. https://qz.com/africa/2161960/gsma-70-percent-of-the-worlds-1-trillion-mobile-money-market-is-in-africa

Omoruyi O (2023) Nigeria's mobile gender gap: 27% of female smartphone owners do not use the internet. https://technext24.com/2023/06/08/mobile-gender-gap-in-nigeria-2023/

OXFAM International (2022) Why the majority of the world's poor are women. Available at https://www.oxfam.org/en/why-majority-worlds-poor-are-women

Ozili PK (2018) Impact of digital finance on financial inclusion and stability. Borsa Istanbul Rev 18(4):329–340. https://www.sciencedirect.com/science/article/pii/S2214845017301503

Porfido D, Marks Z (2020) "Women and the digital economy in Africa" in "women and the changing face of entrepreneurship in Africa". Policy brief. https://africa.harvard.edu/files/african-studies/files/women_entrepreneurship_in_africa_policy_brief_-_digital_economy_final.pdf

Puliti R (2022) Putting Africa on the path to universal electricity access. World Bank Blogs

Safae L (2020) Digital banking in Africa: origins and development outlook. Available at https://infomineo.com/digital-banking-in-africa-origins-and-development-outlook/

The World Bank. Digital Financial Inclusion (2022). Available at https://www.worldbank.org/en/topic/financialinclusion/publication/digital-financial-inclusion

Toesland Finbarr (2021) African countries embracing biometrics, digital IDs. https://www.un.org/africarenewal/magazine/february-2021/african-countries-embracing-biometrics-digital-ids

UNESCO (2021) Women in Africa join forces to promote women's digital skills and entrepreneurship across the continent. https://www.unesco.org/en/articles/unesco-and-women-africa-join-forces-promote-womens-digital-skills-and-entrepreneurship-across

Wikipedia (2023) Internet in Africa. Available at https://en.wikipedia.org/wiki/Internet_in_Africa

World Economic Forum (2022) Women take the lead in Africa's fintech boom. Available at https://www.weforum.org/agenda/2022/03/tech-women-entrepreneurs-africa-fintech/

11 Gender and Digital Financial Inclusion: Assessment of Rural/Urban Dichotomy in Microfinance and Table Banking Operations in ECOWAS

Kemi Funlayo Akeju

Abstract

Various national financial inclusion strategies (NFIS) and measures have been put in place by the government of Economic Community of West African States (ECOWAS) countries for promoting financial service access and affordability for all with most strategies emphasising the benefit of digital savings and payment in enhancing gender equality and women empowerment. Recent activities of digital finance operators through fintechs, blockchain, agent network managers, and technology service providers have been recognised in driving digital financial inclusion. Traditional microfinance and table banking operators focusing on the rural dwellers and the women's society have embraced digital financing, and despite significant strides made, systemic barriers hamper the pace and speed of fostering synergies and harmonisation between digital finance and local financial institutions. As traditional microfinance institutions and operators of table banking in ECOWAS strive to compete with specialised digital finance services operators amidst a competitive market, women and poor households are left behind as they fail to conform to technological advancement. This chapter using selected ECOWAS countries' experiences assesses the challenges, benefits, and risks faced by the traditional microfinance institutions and operators of table banking platforms in their digitalisation process towards enhancing gender equality in financial services. It considers ways of bridging the gender divide in assessing financial services through the partnership of microfinance and table banking operators in rural communities with digital financial services operators given the rise in recognition of the need to include women and the large unbanked society in rural communities. Furthermore, it addresses factors that can advance women's access to digital financial services, build trust in its effectiveness, and promote equality of opportunities in rural areas.

Keywords

Gender financial inclusion · Digital finance · Microfinance institutions · Table banking · ECOWAS

K. F. Akeju (✉)
Department of Economics, Faculty of the Social Sciences, Ekiti State University, Ado-Ekiti, Ekiti State, Nigeria
e-mail: kemi.akeju@eksu.edu.ng

T. A. Ojo (ed.), *Women and Finance in Africa*, Sustainable Development Goals Series,
https://doi.org/10.1007/978-3-031-53337-2_11

Context

As classified by the United Nations,11 Least Developed Countries and 4 lower-middle-income countries made up the ECOWAS region. A large proportion in the region lives in rural areas with limited access to formal financial services. Although progress has been made in providing financial services to the poor in some ECOWAS countries through various digitalization platforms introduced by multinational banks and the microfinance institutions, systemic barriers create constraints to financial services access and reduce the growth pace of financial inclusion, especially for the low-income women in the rural communities.

11.1 Introduction

The development in financial technology (Fintech) has expanded the use of online and digital payments broadly known as digital finance in the delivery of financial market products and services. Digital finance involves the use of technological services by individuals and organisations in their financial transactions, savings, and credit acquisitions through the Internet medium at any period (Ozili 2018; Li et al. 2020). It assists financial institutions to expand their market size and also increase their financial services to rural communities and the wider society and groups. Thus, through a digital transaction platform, firms and individuals can access broad, diversified, affordable, and convenient services offered by financial institutions and other non-financial institutions (Manyika et al. 2016; Fernandes et al. 2021).

Recently, considerable attention is drawn to financial inclusion being the proportion of adults who use financial services through the lens of individuals' availability to gain access to formal financial services (Ulwodi and Peter 2017). The opportunities open to the population of a community of being financially included lie in the strength of their financial sector's capabilities and technological development (Ezzahid and Zakaria 2021). Digital financial inclusion refers to the financial access to technology and informal financial services by the excluded and underserved population (CGAP 2015; Lauer and Lyman 2015; Makina 2019). Revolutionary changes involving fintech through digital and mobile platforms offer new opportunities for the expansion of financial services to households (Demirguc-Kunt et al. 2018; Aziz and Umma 2021). Also, the demands for digital technology applications for financial services access among the youth are on the increase (Purba et al. 2021; Daud et al. 2022), as they easily conform to mobile device operations.

Globally, addressing the financial needs of the financially excluded groups has become one of the major objectives of the United Nations for deepening low- and middle-income countries' financial markets and systems. The dire need for financial inclusion is spurred by the fact that findings from Global Findex data indicate that despite a 7% increase from 2014 to 2017, only 69% of adults across the globe have a financial account in a formal financial institution. Also, about 1.7 billion adults have no financial account globally, either with a financial institution or with a mobile money provider with most living in the developing world, especially in Nigeria, Mexico, Indonesia, Pakistan, Bangladesh, India, Timor-Leste, and Nepal (Demirguc-Kunt et al. 2018).

Reports from the World Bank (2018a, 2021) consider access to financial services to be a key enabler in reducing extreme poverty and boosting shared prosperity. Also, achieving the fifth Sustainable Development Goal – full gender equality in all spheres of life and the empowerment of women and girls is linked to capabilities to access finance and use digital financial services optimally (World Bank 2021). Furthermore, literature on empirical findings revealed that communities with larger financially included people have a lower poverty rate and improve the welfare of the poor (Kast and Pomeranz 2014; Omar and Kazuo 2020), have higher incomes, and contribute to poverty alleviation (Bruhn and Love 2014), financial stability on savings and consumption (Akeju 2022), while others affirmed that the prevailing poverty in less developed economies is directly linked to financial exclusion (N'dri Lasme and Kakinaka 2020).

In Africa, various strategies that align Information and communication technologies (ICT) global practices with the SDGs towards promoting a financial and sustainable environment for the people have been implemented, for example, the African Union (AU) digital transformation strategy was initiated to eliminate the rural/urban digital divide and also bridge the digital divide across gender (African Union 2020). Benefits of digital transformation have been found in West Africa as it creates opportunities to address the challenges of employment and financial inclusion, provides job opportunities across different sectors, and also offers the reduction of transactional costs associated with the transfer of money. Furthermore, the COVID-19 health crisis has shown that digital transformation has the potential to provide innovative solutions in many vital sectors as it aids financial networking. Digitalisation activities rely on the efficient use of communications infrastructure, Internet services, and viable 4G coverage rate which have low supplies in Africa (African Union Commission/OECD 2021). Despite recent growth in digital innovations, African countries are still hampered by poor digital infrastructure, erratic power supplies, and low ICT skills which have a large effect in the rural areas than the urban communities. Despite all the potential of digital finance to promote financial inclusion in West African countries, the adoption of digital financial platforms and mobile money is still concentrated in a few markets while regulations and activities of agent network operators hamper the growth and widespread of financial digitalisation.

Globally, there is a gender gap in financial inclusion; literature on financial inclusion reinforces that women continue to be underserved, with low access to formal financial services and products (Gammage et al. 2017). The importance of promoting gender equality in financial inclusion has been attested to as empirical evidence has shown the influence of financial inclusion on women's empowerment through the control of their earnings and the management of productive activities (Alam 2012; Islam et al. 2014), capabilities to grow their businesses, raise their productivity and earnings, and reduce their chances of being poor (Suri and Jack 2016; Swamy 2014). On marital issues, they have a higher chance of taking decisions on abusive relationships and exposure to intimate partner violence (Pradeep 2014) and have more control and autonomy over their decisions on employment and reproductive issues such as the decision to use contraceptive and family planning (Aker et al. 2016; Suri and Jack 2016; Khalaf and Nahil 2020).

Financial institutions' use of digital finance technologies to serve a wide population creates competition between nationwide corporate banks and small local microfinance/table banking operators with the corporate banks having a higher advantage of capital and digitalisation tools. As payment platforms have shifted from traditional "brick-and-mortar" bank branches and automatic teller machines (ATMs) to digital platforms (Varga 2018; Agarwal and Jian 2020), investment in digital infrastructures such as mobile banking, Internet banking, and point-of-sale machines has supported remote access to digital financial services, thereby promoting financial inclusion and reducing banks' operating costs (Chen and Xiaohui 2021).

Microfinance institutions (MFIs) have social and financial objectives centring on the reduction of poverty by providing financial services to households excluded from the formal financial system and also the generation of profit for stability. Some microfinance operators have enlarged their operations using digital platforms through financial agencies capturing the rural dwellers and the women's society. Although remarkable significant strides have been made, systemic barriers hamper the pace and speed of fostering synergies and harmonisation between digital finance and local financial institutions (Dang and Huong 2020). A large number of women especially those not literate and living in rural areas face the challenge of accessing and using digital financial services, as they are not using digital mobile communications networks. Generally, technical knowledge and desirability are critical factors in determining digital financial inclusion while unequal social relations and opportunities create gender-specific barriers to financial access, which enhance persistent gaps and inequalities (Fouejieu et al. 2020).

11.2 Microfinance and Table Banking Paradigms: Operations, Benefits, and Challenges of Digitalisation in ECOWAS

Microfinance is associated with the generation of microcredit for low-income households. Over time, it includes access to high-quality and affordable financial services for low-income households and rural communities to finance their production activities, consumption, savings, insurance, payments, and remittances (Halimah 2017). It has become popular among rural communities, women, and generally within the informal sector, and it is believed that good microfinance operations reduce poverty as it captures the delivery of financial services such as credit and savings to those without access to formal financial institutions (Morduch 2000). Microfinance banks are financial institutions that are considered a relatively new strategy to combat poverty in any country (Churchill 2020), due to the challenges associated with credit facilities offered by formal financial institutions to local communities. Microfinance institutions (MFIs) are one of the major policy instruments that play significant roles in poverty reduction activities, particularly in enabling poor people generate their income (Odoom et al. 2019; Khan et al. 2021). Microfinance institution (MFI) operations largely target the poor (especially poor women) and operators of small businesses to grant them credits and small loans with minimum interest rates and ensure their capability to save.

Moreover, microfinance and table banking operations have helped entrepreneurs by granting credits and small loans to them with minimum interest rates. It also ensures they benefit from financially related services they are unable to access from commercial banks due to the high transactional costs (Mogaji et al. 2021), and it is generally used to provide help to micro- and small-scale entrepreneurs and companies which generally have difficulties in reaching financial and banking services due to the cost and financial resource limitations (Ruth and Jagongo 2018; Fieve and Chrysostome 2024). Various empirical evidence reveal that microfinance institutions have made a positive impact on poverty reduction and improved individuals' circumstances, with evidence that microfinance can help the rural poor improve their economic conditions and also help customers to increase their income through the accessibility of microcredit (Oshinowo et al. 2018; Prior and Toni 2019).

As digital financial services promote financial inclusion, microfinance and table banking operators are embracing digital finance with most incorporating the use of mobile devices and partnering with a digital agency or financial network. Digital technology is ultimately changing the microfinance value chain; these have enabled them to offer financial services using automatic teller machine (ATMs) services, digital payments, and money transfers. With the rapid advancement in the use of mobile phones, electronic apps, and devices, market women benefit from microfinance through their partnerships with digital agencies, and financial platform operators offer great potential to deepen financial inclusion. The entry of microfinance service providers into the digital economy is subject to the size of their customers, groups of networks, and access to a licensed mobile phone operator. The product or service distributed (mobile money, mobile banking) belongs to mobile phone operators or banks.

11.3 Digitalisation of Microfinance Bank's Operations in Nigeria

Microfinance emanates from the establishment of the community banks in 1991 as part of the efforts and drives for rural development by the government. Despite these efforts, community banks were reported not capable of achieving their objectives, hence the need for the establishment of regulated microfinance banks in 2005 (Moruf 2013). Various microfinance institutions are presently operational in Nigeria promoting financial inclusion by helping individuals, farm owners, and small- and medium-scale business owners to access loans to start and enlarge their enterprises. Although their operations are similar to the commercial banks, their presence in the rural communities has helped rural dwellers to make progress in their businesses. With recent technological advancements in the banking sector, many microfinance banks in Nigeria are adopting digital innovative ways of doing business, offering banking services at little or no cost to their customers.

At present, not all microfinance banks operate digital financing due to their traditional practices and the challenges of access to infrastructures, but a good number of digital microfinance operations have become notable in Nigeria. A good example is "KUDA Microfinance Bank", a digital microfinance bank founded in 2018 with a full licence from the Central Bank of Nigeria to help Nigerian business owners' access and easy financial operations with minimum operational/transactional fees. Customer's savings are safe and secured with deposits insured by the Nigerian Deposit Insurance Corporations (NDIC). It offers digital credit and debit alerts with debit cards across ATMs across the country, with the encouragement of saving habits and budgeting and the ability to access short-term loans and Kuda overdraft. Kuda's operation helps women in entrepreneurship to effectively have access to banking services without excessive charges, manage their accounts, and help promote a savings culture through the provision of a 15% interest on an annual savings rate. Another notable digital microfinance bank is "V Bank", a platform of VFD Microfinance Bank with non-transactional charges and maintenance fees which also offers customers interest on their savings every month. Other digital microfinance banks operational in Nigeria include Lapo, Mutual Trust, AB microfinance, etc. with most of them using mobile application stores or Google Play stores, making it easy for customers to connect and commence operation within a few minutes of connectivity.

11.4 Microfinance Institutions and Digitalisation in Ghana

Various deposit-taking microfinance institutions are operational in Ghana. These deposit-taking microfinance institutions have the institutional structure and regulatory approval required to mobilise intermediate deposits. In 2019, the Bank of Ghana had the licence of about 347 microfinance firms revoked, allowing only financially solvent finance institutions operational which are licensed and regulated as banks or operate under a special category for deposit-taking microfinance institutions (MFIs) created by the regulatory authorities (Akolgo 2022). Microfinance operations spread rapidly in Ghana in the late 2000s with operational guidelines establishing four tiers of microfinance institutions issued in 2011: Tier 1 (deposit-taking microfinance institutions) undertaken by Rural and Community Banks, Finance Houses, and Savings and Loans Companies which are supervised by the Bank of Ghana, Tier 2 (deposit-taking microfinance institutions) undertaken by licensed microfinance companies, credit unions, and other financial service providers that are deposit-taking and profit-making whose operations are monitored and supervised by the Bank of Ghana and credit unions regulated by Ghana Cooperative Credit Unions Association, Tier 3 (non-deposit-taking microfinance institutions) operated by money lenders and non-deposit-taking financial non-governmental organisations, Tier 4 (financial activities of individual Susu collectors and individual money lenders) (Duho et al. 2023).

The National Financial Inclusion and Development Strategy (NFIDS) 2017–2023 plan has been adopted by the Government of Ghana for ensuring the adoption of digital financial services and their optimum use (Fiador and Amidu 2021). Examples of operational digital microfinance institutions in Ghana are 3AS Microfinance Limited, Baobab Microfinance Company, Wema Microfinance Limited, etc. Microfinance institutions have lower minimum capital requirements and cannot provide all the services international commercial banks can offer. The ability of microfinance finance institutions to accept deposits contributes to financial inclusion by providing clients with a secure place to save and by facilitating savings; these institutions better respond to client needs and can expand services to non-borrowers. Accepting deposits also enhances the MFI's sustainability and places it in a better position to access commercial sources of funding.

11.5 Microfinance Operations in Senegal

As a member of the West African Economic and Monetary Union (WAEMU), Senegal's financial and macroeconomic policies are designed and implemented by the Union. The recently launched Senegal national strategy for gender equality "SNEEG 2016 – 2026" identifies the need to raise women's access to financial services for economic empowerment (New Faces New Voices 2020). The financial inclusion strategies have been promoted with recent digital transformation which has changed the over-the-counter banking services operational in Senegal with the rise of mobile money operators – "Tigo and Orange Money" offer customers' electronic money accounts which aid electronic transfer payments.

The microfinance known as "Services Financiers Descentralisés" operations has been strengthened with digital platforms with many capturing the rural and low-income earners. Examples of microfinance institutions with digitalised operations are the Baobab microfinance formerly called Microcred, Carrie microfinance, and Fansoto social microfinance. Baobab has many subsidiaries across African countries which operate an agency banking model with a secure web-based agency banking platform (developed by Software Group) and a mobile application running on tablets for agents. Fansoto microfinance captures entrepreneurs who are mostly women operating in low- and middle-scale industries. Microfinance institutions in Senegal provide grants in the form of microcredits to entrepreneurs in mostly agriculture and trading business and contribute largely to social development in Senegal. Benefits of microfinance operations in raising the savings habits of female entrepreneurs have been identified (Scanlon et al. 2019), and the challenges of digitalisation in microfinance institutions have been identified as the ability to design a new product for customers, networking, infrastructure, high cost of Internet subscription, and poor connectivity, especially in rural areas.

11.6 Operations of Microfinance Institutions in the Gambia

In the Gambia, microfinance institutions manage and facilitate their customers' financial transactions such as deposits, loan repayment, and renewal. They are categorised into (i) village savings and credit associations which are community-based financial institutions managed by members, (ii) finance companies owned by individuals with a regularised minimum capital as shares, and (iii) Fiduciary financial institutions with the operational licence granted for their activities. Although at present, a large portion of the financial institutions in the Gambia still operates manual documentation, processing, and over-the-counter system, information technology has provided various digital financial products which have revolutionised the delivery of financial services to entrepreneurs, market women, and a large population that were without formal account ownership in formal banking institutions. The use of mobile phones for financial transactions known as mobile money has become popular while microfinance institutions with digital platforms have increased, but these are inhibited by sociocultural, finance, knowledge, and skills.

11.7 Rural/Urban Dichotomy of Microfinance in Financial Inclusion

The functioning of microfinance institutions in the promotion of financial inclusion through the strengthening of savings and credit creation for low-income and business operators in ECOWAS are rendered in distinct disparity across urban and rural communities (Madichie et al. 2021). Although recent development across the financial sector on the need to address poverty at the rural level has strengthened microfinance institutions with its benefits outweighing their weaknesses, there are still challenges facing it which often promote the rural-urban divide. These challenges include the inaccessibility of capital to foster loan portfolios which largely affect microfinance branches in the rural areas than in the urban centres and limit their capacity for expansion and coverage of a wide network. Also, the strong dichotomy between rural and urban areas' digital financing is linked to the supply of ICT and infrastructure with the rural areas often at a disadvantage due to poor Internet and digital technology availability (Isley and Low 2022). In West Africa, the rate of access to electricity is 52% on average, with power outages that can reach only 80 h per month (World Bank 2018a).

In Ghana, a large proportion of those without access to financial accounts live in rural areas, providing access to financial products and services by institutions and banking agencies at an affordable and fair cost involves digital networking which is not adequately made available to rural and remote communities (Gyasi et al. 2019). Financial institutions and banks are needed to provide financial services to the poor and disadvantaged social groups in rural and remote communities to reduce the demand and supply gap in financial services. Most entrepreneurs in rural communities are women in need of financial management and help to build their savings and payment systems to promote their business and achieve greater financial stability over the long term.

Generally, microfinance goals focus on the empowerment of the poorest financially, especially among women participating in the small-scale business sectors through enhancing their control of financial matters, fostering savings capacity and budgetary tool management. Microfinance institutions using traditional and conventional counter and table banking operations have helped improve the savings strategies with one-on-one interactions with women in remote rural communities who are without any form of financial and digital literacy. With the introduction of digital platforms, these goals are not accomplished among the large group of people in rural areas who are not digitally compliant and cannot easily adjust to the use of mobile phones for digital operations. Also, the operations of table banking mostly by women entrepreneurs through the accumulation of funds by members saving which are disbursed immediately from the table as short-term or long-term loans has strengthened and improved their investment decisions.

Arguments on the negative implications of digitalisation on the traditional value of microfinancing in rural areas have emerged with the claim that digitalisation removes the interaction between the microfinance agent and customers, obstructs the training, and supports market benefits women receive from microfinance agents (Ozili 2020; Siwale and Godfroid 2022). Also, the relational association between microfinance officers and their customers often promote trust and confidence which enhance saving and the lending system (Behr et al. 2020), thus raising the call for a blended operation of the traditional system of microfinance operations with the digitalisation system.

11.8 Gender and Financial Digitalisation

Sustainable Development Goal 5 identifies the need to strengthen women. Recognising gender in microfinance involves not only targeting a programme for women but recognising the position of women in the society and daily lives and also considers providing financial support for women to overcome the hurdles associated with lives and help achieve their goals and aspirations (D'espallier et al. 2013). In addition to its economic benefits, women's financial inclusion promotes greater equality and societal well-being as recent research provides solid evidence of women using their earned income and savings more productively than men, channelling a large share to children's nutrition, clothing, health, and education.

There remains a need for affordable financial services, including among vulnerable, disadvantaged, and low-income groups and individuals (Saatcioglu and Corus 2014; Agyemang-Badu et al. 2018). Evidence of disparity in choice of financial accounts operations has also been reported with a higher percentage of men having formal financial accounts at financial institutions than women, while women's savings and borrowing in informal financial systems are higher. These results have been attributed to the operating system and the control of income in the households. Cultural and social differences also play a challenging role; attitudes and opinions of men and women differ in repayment of loans as most women tend to comply with repayment policies as they tend to be ashamed of loan default while most men are indifferent to it. Women often lack control over cash management within their household and may be dependent on their husbands to access financial services. Women's financial inclusion has been an underused source of growth that if addressed can contribute to sustainable and inclusive development. Given the important role women play in shaping the next generation, their financial inclusion will lead to greater security and prosperity for women and men, their families, enterprises, and communities.

Table 11.1 Percentage of age 15+ with account ownership in 2017 (Findex, 2018 report)

S/N	Countries/group	Women-country data (%)	All adult
1	Benin	28.6	38.5
2	Burkina Faso	34.5	43.2
3	Cote dívoire	35.6	41.3
4	Ghana	53.7	57.7
5	Guinea	19.7	23.5
6	Liberia	28.2	35.7
7	Mali	25.7	35.4
8	Niger	10.9	15.5
9	Nigeria	27.3	39.7
10	Senegal	38.4	42.3
11	Sierra Leone	15.4	19.8
12	Togo	37.6	45.3
	Sub-Sahara Africa	36.9	42.6
	World	64.8	68.5

Source: Extracted and compiled by the Author from: (The Little Data Book on Financial Inclusion 2018) (World Bank 2018b)

Although significant progress has been made in enhancing financial inclusion around the world, financial gender gaps known as the disproportionate exclusion of women from access to and usage of financial services remain persistent globally, especially in West African countries. Evidence of the gender divide in account ownership in West African countries is found in Table 11.1, where a report on all adults is higher than women in all countries reported and also in comparison with aggregate data on the sub-Saharan African region and the World at large.

The data presented in Table 11.1 shows that women's account ownership in most West African countries is lower than that of the sub-Saharan African region and the world at large with only Ghana (53.7%) higher than the sub-Saharan African regional aggregate. Generally, the gender divide challenges are compounded by three major groups of barriers: (i) lack of access to the basic requirements for financial inclusion, (ii) unequal laws, norms, and cultural and regulations barriers, and (iii) Inadequate digital financial infrastructure. However, there continue to be gender-based constraints to women's full financial inclusion that also extend to digital systems and explain the persistent gender gaps despite the digital expansion, and among these barriers are women's financial literacy and access to mobile technology (Tay et al. 2022).

Digital financial services appear to offer innovative platforms to address financial exclusion and have the potential to overcome restrictions brought on by geography, reduce the transaction costs of using financial services, enhance transparency, and therefore increase trust in financial systems. Through data capturing and storage, digital platforms can reduce the informational asymmetries between lenders and borrowers that keep large segments of mostly poor populations excluded from financial systems. They can also improve financial autonomy and privacy for users and potentially reduce the risk of co-option and appropriation of these funds by others. The global revolution in mobile communications and rapid advances in digital payment systems offer a route to connecting the poor to financial services and products. These digital platforms can contribute to closing the existing gender gap in financial inclusion, affecting both the demand- and supply-side factors that underlie women's current exclusion from financial systems.

11.9 Conclusion

Africans must consider the use of financial digitalisation to enhance women's empowerment and create an avenue to raise their financial strength and eradicate poverty. The digitalisation of microfinance banks is inevitable with rising competition among financial institutions and the need to include rural communities in digital financial services. Not only does it promote the use of the financial account and grant outreach to rural areas and low-income clients, but it also enhances small and medium enterprises' opportunity to have access to financing and increases business offers and revenue for microfinance operators. This is done through the partnerships of microfinance institutions with providers such as multinational organisations, commercial banks, and fintech companies using mobile platforms and Internet networks. Digital solutions can help financial institutions deepen customer engagement, increase product usage, and promote financial inclusion as a large proportion of the West African population is financially excluded.

For West African countries to take full advantage of the benefits of digital transformation requires overcoming regional barriers and challenges of financial education, gender divide, electricity supply, cyber corruption, Internet scam operations, quality communications infrastructure, and improvement in the legal and regulatory framework. Digital advancement in microfinance banks in West Africa faces the challenges of service disruptions which emanate from inadequate systemic operations, insecurity and instability, poor telecommunication services, lack of investment, and low capital, particularly in remote areas; also, the impact of political instability on economic growth implies the drive to digital access and infrastructure development. Moreover, operational challenges in the management of agent networks to ensure adequate compliance and efficiency are higher, particularly in remote areas. Transformative solutions to ensure adequate digitalisation of microfinance institutions and table banking in West Africa should consider drastic improvements in the financial system to increase financial inclusion through the availability of smart devices such as smartphones which can be supplied to customers for use in payment services.

Globally, there is a gap in financial inclusion between men and women, and bridging this gap is imperative considering the impact of women's financial inclusion on poverty eradication, household socioeconomic development, and attainment of related Sustainable Development Goals. The expected benefits of digital financial inclusion by microfinance institutions can be fully realised if training on the use of digital platforms such as mobile phones, personal computers, and related devices is conducted for market women and entrepreneurs with low-level education to ensure the widespread of digital platforms for financial activities. Microfinance institutions should disaggregate their customer's data by gender, tailor their products (financial and value-added) to meet women's needs, and consider the suitable financial opportunities and schemes for women-led group entrepreneurs through market research.

Nonetheless, the interactional linkage between microfinance officers and their customers need to be kept functioning through the traditional model of microfinance operating system so as not to marginalise the poor and the large population without access and knowledge of digitalisation.

References

African Union (2020) The digital transformation strategy for Africa (2020–2030). Retrieved from African Union. https://au.int/sites/default/files/documents/38507-doc-dts-english.pdf

African Union Commission/OECD (2021) Digital transformation for youth employment and agenda 2063 in West Africa. In: Africa's development dynamics 2021: digital transformation for quality jobs. African Union Commission/OECD Publishing, Addis Ababa/Paris. https://doi.org/10.1787/bb1d4e48-en

Agarwal S, Jian Z (2020) FinTech, lending and payment innovation: a review. Asia-Pacific J Financ Stud 49(3):353–367

Agyemang-Badu AA, Agyei K, Duah EK (2018) Financial inclusion, poverty and income inequality: evidence from Africa. Spiritan Int J Poverty Studies 2(2):1–18

Akeju KF (2022) Household financial behaviour: the role of financial inclusion instruments in Nigeria. J Sustain Financ Invest:1–13

Aker J, Rachid B, Amanda MC, Niall T (2016) Payment mechanisms and antipoverty programs: evidence from a mobile money cash transfer experiment in Niger. Econ Dev Cult Chang 65(1):1–37

Akolgo IA (2022) Collapsing banks and the cost of finance capitalism in Ghana. Rev Afr Polit Econ 49(174):624–633

Alam S (2012) The effect of gender-based returns to borrowing on intra-household resource allocation in rural Bangladesh. World Dev 40(6):1164–1180

Aziz A, Umma N (2021) Rethinking digital financial inclusion: evidence from Bangladesh. Technol Soc 64:101509

Behr P, Drexler A, Gropp R, Guettler A (2020) Financial incentives and loan officer behavior: multitasking and allocation of effort under an incomplete contract. J Financ Quant Anal 55(4):1243–1267

Bruhn M, Love I (2014) The real impact of improved access to finance: evidence from Mexico. J Financ 69(3):1347–1376

CGAP (2015) Digital financial inclusion: implications for customers, regulators, supervisors, and standard-setting bodies. CGAP brief. https://www.cgap.org/sites/default/files/researches/documents/Brief-Digital-Financial-Inclusion-Feb-2015.pdf

Chen W, Xiaohui Y (2021) Financial inclusion in China: an overview. Front Bus Res China 15(1):1–21

Churchill SA (2020) Microfinance financial sustainability and outreach: is there a trade-off? Empir Econ 59(3):1329–1350

D'espallier B, Guerin I, Mersland R (2013) Focus on women in microfinance institutions. J Dev Stud 49(5):589–608

Dang TT, Huong QVU (2020) Fintech in microfinance: a new direction for microfinance institutions in Vietnam. J Bus Econ Environ Stud 10(3):13–22

Daud I, Nurjannah D, Mohyi A, Ambarwati T et al (2022) The effect of digital marketing, digital finance and digital payment on finance performance of Indonesian SMES. Int J Data Netw 6(1):37–44

Demirguc-Kunt A, Leora K, Dorothe S, Saniya A (2018) The global findex database 2017: measuring financial inclusion and the fintech revolution. World Bank Publications

Duho TKC, Duho DM, Forson JA (2023) Impact of income diversification strategy on credit risk and market risk among microfinance institutions. J Econ Adm Sci 39(2):523–546

Ezzahid E, Zakaria E (2021) Financial inclusion, mobile banking, informal finance and financial exclusion: micro-level evidence from Morocco. Int J Soc Econ 48:1060–1086

Fernandes C, Rosa BM, Jorge C (2021) The contribution of digital financial services to financial inclusion in Mozambique: an ARDL model approach. Appl Econ 53(3):400–409

Fiador VOL, Amidu M (2021) Inclusive finance, bank pricing behaviour, and livelihood activities of households in Ghana. SN Bus Econ 1(1):1–29

Fieve JKD, Chrysostome EV (2024) Credit cooperative lending loans as challenges and opportunities for women entrepreneurship in Africa: evidence from Ghana. J Afr Bus 25(1):94–114

Fouejieu A, Ratna S, Martin C, Shiyuan C (2020) Financial inclusion and inequality: a cross-country analysis. J Int Trade Econ Dev 29(8):1018–1048

Gammage S, Kes A, Winograd L, Sultana N, Hiller S, Bourgault S (2017) Gender and digital financial inclusion: what do we know and what do we need to know. International Center for Research on Women (ICRW)

Gyasi RM, Adam AM, Phillips DR (2019) Financial inclusion, health-seeking behavior, and health outcomes among older adults in Ghana. Res Aging 41(8):794–820

Halimah AM (2017) Impact of microcredit scheme on the development of entrepreneurship skills among low income households in a developing country context. Int J Econ Financ Issues 7(1):583–593

Islam MS, Faysal AM, Muhammad SA (2014) The role of microcredit program on women empowerment: empirical evidence from rural Bangladesh. Dev Country Stud 4(5):90–97

Isley C, Low SA (2022) Broadband adoption and availability: impacts on rural employment during COVID-19. Telecommun Policy 46(7):102310

Kast F, Pomeranz D (2014) Saving more to borrow less: experimental evidence from access to formal savings accounts in Chile. No. w20239 (National Bureau of Economic Research)

Khalaf LS, Nahil IS (2020) Microfinance: an emerging tool for women's economic empowerment. In: Entrepreneurial opportunities. Emerald Publishing Limited

Khan AA, Khan SU, Fahad S, Ali MAS, Khan A, Luo J (2021) Microfinance and poverty reduction: new evidence from Pakistan. *International Journal of Finance & Economics* 26(3):4723–4733

Lauer K, Lyman T (2015) Digital financial inclusion: Implications for customers, regulators, supervisors, and standard-setting bodies. No. 95210. (The World Bank, 2015)

Li J, Wu Y, Xiao JJ (2020) The impact of digital finance on household consumption: evidence from China. Econ Model 86:317–326

Madichie NO, Bolat E, Taura N (2021) Digital transformation in West Africa: a two country, two-sector analysis. J Enterp Commun People Places Glob Econ 15(2):246–257

Makina D (2019) The potential of FinTech in enabling financial inclusion. In: Extending financial inclusion in Africa. Academic, London, pp 299–318

Manyika J, Susan L, Marc S, Olivia W, Chris B (2016) Digital finance for all: powering inclusive growth in emerging economies. McKinsey Global Institute, pp 1–15

Mogaji E, Adeola O, Hinson RE, Nguyen NP, Nwoba AC, Soetan TO (2021) Marketing bank services to financially vulnerable customers: evidence from an emerging economy. Int J Bank Mark 39(3):402–428

Morduch J (2000) The microfinance schism. World Dev 28(4):617–629

Moruf O (2013) Evaluation of the Nigerian microfinance banks credit administration on small and medium scale enterprises operations. Int Rev Manag Bus Res 2(2):505–517

N'dri Lasme M, Kakinaka M (2020) Financial inclusion, mobile money, and individual welfare: the case of Burkina Faso. Telecommun Policy 44(3):101926

New Faces New Voices (2020) Women's financial inclusion in Senegal

Odoom D, Fosu KO, Ankomah K, Amofa MB (2019) Exploring the contributions of microfinance institutions to the Ghanaian economy: a study at Takoradi. J Econ Sustain Dev 10(1):77–90

Omar AM, Kazuo I (2020) Does financial inclusion reduce poverty and income inequality in developing countries? A panel data analysis. J Econ Struct 9(1):1–25

Oshinowo O, Olayide O, Azeez M (2018) Effect of microfinance institutions on rural households' well-being in Oyo State: case study of IFAD/RUFIN supported project. Afr J Sustain Dev 8(1):67–85

Ozili PK (2018) Impact of digital finance on financial inclusion and stability. Borsa Istanbul Rev 18:329–340

Ozili PK (2020) Contesting digital finance for the poor. Digital policy, regulation and governance 22(2):135–151

Pradeep KP (2014) Role of micro finance in reduction of domestic violence against women: an economic analysis. Indian J Econ Bus 13(3):449

Prior F, Toni M (2019) Quantitative study on the impact of branchless banking on microfinance institutions. Ann Public Coop Econ 90(4):641–668

Purba J, Samuel S, Budiono S (2021) Collaboration of digital payment usage decision in COVID-19 pandemic situation: evidence from Indonesia. Int J Data Netw Sci 5:557–568

Ruth IO, Jagongo A (2018) Microfinance services and financial performance of small and medium enterprises of youth SMEs in Kisumu County, Kenya. Int Acad J Econ Financ 3(1):24–43

Saatcioglu B, Corus C (2014) Poverty and intersectionality: a multidimensional look into the lives of the impoverished. J Macromark 34(2):122–132

Scanlon CC, Scanlon K, Scanlon T (2019) The influence of collectivism on microfinance in Senegal. In: Entrepreneurship and development in the 21st century. Emerald Publishing Limited

Siwale J, Godfroid C (2022) Digitising microfinance: on the route to losing the traditional 'human face' of microfinance institutions. Oxford Dev Stud 50(2):177–191

Suri T, Jack W (2016) The long-run poverty and gender impacts of mobile money. Science 354(6317):1288–1292

Swamy V (2014) Financial inclusion, gender dimension, and economic impact on poor households. World Dev 56:1–15

Tay L-Y, Tai H-T, Tan G-S (2022) Digital financial inclusion: a gateway to sustainable development. Heliyon 8(6):e09766.

Ulwodi D, Peter M (2017) Barriers of financial inclusion in Sub-Saharan Africa. J Econ Sustain Dev 8(14):66–81

Varga D (2018) Fintech: supporting sustainable development by disrupting finance. Budapest Manag Rev 8(11):231–249

World Bank (2018a) Poverty and shared prosperity 2018: piecing together the poverty puzzle. World Bank, Washington, DC

World Bank (2018b), The Little Data Book on Financial Inclusion

World Bank (2021) The drive for financial inclusion: lessons of World Bank Group experience. 30 June 2021

12 Gendered Dimension for Digital Financial Inclusion in Nigeria

Blessing Olanrewaju and Yetunde A. Aluko

Abstract

Financial inclusion offers a myriad of benefits for economic development. Accordingly, it has been incorporated into national, regional and international development objectives. However, increasing gender disparities, especially in terms of access to and delivery of financial services, is counterproductive to the achievement of the United Nations Sustainable Development Goals, which emphasizes leaving no one behind by 2030. Several strategies and efforts have been put in place to close the financial gap between men and women in developing countries, but with little success. Given the vital role digital finance plays in advancing financial inclusion, this chapter examined the gender gap in access to financial services and the role of digital finance in addressing this gap in Nigeria. The findings confirmed the pervasiveness of gender gaps in access to digital financial services in Nigeria as more men than women have access to digital finance platforms. This may be attributed to low levels of digital literacy, inadequate financial infrastructure and low Internet penetration, among others. On the whole, the chapter concludes that financial inclusion seems like a mirage in Nigeria, but digital finance has great potential to encourage financial inclusion in the country since it has enhanced access to financial products and services by gender. Therefore, the chapter advocates the need to adequately equip the Nigerian financial sector to provide improved financial services to both women and men, without prejudice or discrimination and build physical infrastructure to foster financial integration in Nigeria.

Keywords

Financial inclusion · Digital finance · Gender gap · Financial products · Nigeria

12.1 Background and the Problem

The pursuit of financial inclusion has been greatly stimulated over the past decades due to its potential impact on inclusive economic development. Literature has shown that financial inclusion is a growth enabler, as it stimulates the mobilization of domestic resources, fosters employment generation, alleviates poverty, and promotes income

B. Olanrewaju
Economic and Business Policy Department, Nigerian Institute of Social and Economic Research (NISER), Ibadan, Oyo State, Nigeria

Y. A. Aluko (✉)
Social Policy Department, Nigerian Institute of Social and Economic Research (NISER), Ibadan, Oyo State, Nigeria

T. A. Ojo (ed.), *Women and Finance in Africa*, Sustainable Development Goals Series,
https://doi.org/10.1007/978-3-031-53337-2_12

equality (Omar and Inaba 2020; Riley et al. 2020). However, the pervasive financial gender gap in most developing countries, including Nigeria, could reduce the dividend from inclusive financial services and potentially slow economic growth.

Several international and national efforts have been undertaken to address the gender gap in financial inclusion. These include the United Nations (UN) Sustainable Development Goal (SDG) 5, the Alliance for Financial Inclusion (AFI) network, and the National Financial Inclusion Strategy (NFIS). In particular, one of the targets of SDG 5 is to promote equal rights to financial services, especially among women. Also, during the 2016 global policy forum, the Alliance for Financial Inclusion (AFI) made a public statement to bridge the gender gap in financial inclusion through the implementation of the Denarau Action Plan that recommends closing the global financial gender gap by 2021, and which Nigeria is committed to.

In Nigeria, the National Financial Inclusion Strategy (NFIS) aimed to reduce the proportion of the financially excluded population from 46.3% in 2010 to 20% in 2020. Also, under the NFIS, gender considerations have been adopted in the development of financial products and services to improve women's access to financial services. This implies that closing the gender gap in the access to and usage of financial services is imperative to the achievement of the United Nation's laudable goals as well as for national and global development.

Unfortunately, despite concerted efforts to narrow the gender gap in financial inclusion, the situation remains the same, particularly in most developing regions of the world. For instance, the World Bank's Global Findex data revealed that in 2017, while 71% of the world's male population possesses a financial institution account, only 64% of the female populace owned a formal financial institution account, indicating a margin of 7% (Demirgüç-Kunt et al. 2018). The situation is worse in sub-Saharan Africa where only 27% of female possess a financial institution account in 2017 against 38% recorded for male (Demirgüç-Kunt et al. 2018). In Nigeria, only 39% of the total population owned an account with a financial institution in 2017, and the gender gap in financial inclusion has continued to grow over the years (Aluko 2018/2019).

According to the Alliance for Financial Inclusion [AFI] (2017), the main barriers to women's financial inclusion in Nigeria are primarily demand driven. Based on the outcome of the 2010 Enhancing Financial Innovation and Access (EFInA) survey, Nigerian women are financially excluded due to inadequate income, lack of access to or trust in financial services, irregular income, financial literacy, unemployment and eligibility for and affordability of financial services, amongst others (AFI 2017). In spite of these challenges, it is interesting to note that recent advancement in technology and globalization has engendered the development of various digitally driven financial services and products that facilitates access to and use of finance as well as the process of financial intermediation.

This innovative initiative, in the form of digital Financial Services (DFSs), has been identified as a platform through which the gender gap in financial inclusion can be closed (AFI 2017). In a nutshell, DFSs refer to the digital platforms through which financial transactions take place. A myriad of financial services such as payments, savings, credit, remittances and insurance can be offered and transacted via digital platforms, thereby eliminating the need for the traditional mode of banking and enhancing access to and usage of financial services. However, the main components of DFSs include mobile banking, mobile wallets and mobile money and transfers (Gammage et al. 2017).

It is noteworthy that high-income countries with high uptake of digital financial services have a high financial inclusion rate and lesser gender gap in financial inclusion than countries with low uptake. For example, 83% of the high-income countries population owned a debit card in 2017 with 84% of this population being male and 82% being female, indicating a gender gap of only 2%. Further, 91% of the total population of high-income countries made or received payments digitally in 2017 with 91% of this population being male and 90% being female, representing a 2% gender gap. Conversely, only 40% of the total inhabitants of developing economies had a debit card in 2017, with 45% male and 35% female, demonstrating a gender gap of 10%. In addition, the proportion of the developing countries population who received or made payments digitally in 2017 is only 44%, of which 48% are male while 39% are female, which denotes a 9% gender gap in financial inclusion.

In general, women have less access to financial products and services, especially in developing countries. The World Bank regards this as a problem because the low access to financial products and services prevents women from making everyday decisions that other gender tend to take for granted (Helms 2006). It also subdues women's political and economic power while limiting their contributions in the economic and political space (Ghoh 2022). Meanwhile, evidence abounds that women have the potential to make meaningful economic and political contributions like their male counterparts (Mckinsey Global Institute 2015). The Global Partnership for Financial Inclusion [GPFI] (2020), however, claimed that the economic power of women can be built through access to and usage of responsible (safe, secured, fair and customer-friendly) digital financial services. Digital financial services can be used to build women's financial resilience, particularly as the world recovers from the devastating COVID-19 pandemic.

The foregoing, therefore, suggests that effective uptake of digital financial products and services is absolutely necessary to bridge the gender gap in financial inclusion across the globe, particularly in developing and emerging economies. Consequently, this chapter examines the role of digital finance in reducing the generalized gender gap in financial inclusion in Nigeria. The chapter demystifies the gender dimension of the access to and usage of digital financial services and products in Nigeria with a view to proffering plausible policy recommendations geared towards closing the gender gap in financial inclusion.

The remaining part of this chapter is structured as follows: Section 12.2 presents a literature review comprising some conceptual clarifications and empirical review of the extant literature; Section 12.3 presents some stylized facts on the gender disparity in access to financial services in Nigeria; Section 12.4 takes cues from countries who have leveraged digital financial services to reduce the gender disparity in financial inclusion and policy priorities; and Sect. 12.5 concludes the study.

12.2 Financial Inclusion and Digital Financial Inclusion: Conceptual Review

12.2.1 Financial Inclusion

The concept of financial inclusion has been broadly defined in the existing literature. However, the summary of these definitions is that financial inclusion basically involves equal access to and usage of financial products and services by all economic agents, irrespective of their age, gender, region or status. The Global Partnership for Financial Inclusion (2016) defines financial inclusion as a situation where all adult labour force can effectively access a number of financial services offered by formal financial institutions including credit, payments, savings, investments and insurance. Broadening the definition has shown that effective access to financial services transcends ownership or access to financial services but involves their use. In addition, ensuring effective access implies that the financial services are convenient to use, affordable to the users, responsibly delivered and meet the demands of the financially excluded and marginalized population.

Financial inclusion appears to be a key prerequisite for economic development as its benefits stem from the application of financial technology and digital financial services such as cards for payments and mobile money services. Specifically, financial inclusion, associated with the use of mobile money devices, is capable of alleviating persistent poverty, enhancing gender parity, reducing transaction costs, boosting savings and fostering effective management of financial risks by households (Ferrata 2019).

12.2.2 Digital Financial Inclusion

Digital financial inclusion (DFI) has been largely conceptualized by various authors. For instance, according to the GPFI (2016), DFI is a general term describing the use of DFSs to promote financial inclusion. Lauer and Lyman (2015) argue that DFI involves the use of cost-effective digital devices to provide financially excluded individuals with a variety of customized and affordable formal financial products and services. According to Naumenkova et al. (2019), DFI facilitates universal access to formal financial services based on the implementation of its key components including a digital interface between financial service providers and customers, digitalization of financial platforms and products, use of digital service channels and application of digital business designs and models for customer service and fund-raising arrangement. Thus, on the basis of its key components, DFI undeniably fosters sustainable financial inclusion and offers a comparative advantage over the traditional approach to financial inclusion.

Digital financial inclusion seeks to provide financial services through digital means to all economic agents, thereby curbing poverty, enhancing financial intermediation and fostering the realization of the Sustainable Development Goals (SDGs) (Ozili 2022). Ferrata (2019) argued that even though none of the SDGs is overtly associated with the role of finance and financial inclusion, the duo is implicitly instrumental to the achievement of the goals as digital technologies could reduce the barriers to finance, thus enhancing people's contributions to the SDGs. Put differently, the availability and accessibility of DFSs, provided by both formal and informal financial institutions, are capable of fostering the financial inclusion of the financially excluded, including women, hence accelerating the achievement of the overarching agenda of the SDGs—leaving no one behind by 2030.

12.2.3 Digital Financial Services (DFSs) and Women Financial Inclusion: Empirical Evidence

The role of digital financial services (DFSs) as a tool for greater financial inclusion has been widely reviewed in the empirical literature. But studies on the gender impact of DFSs, particularly in Nigeria, are rare. Most existing studies on the subject agree that the adoption of DFSs goes a long way towards bridging the gender gap in financial inclusion around the world. For example, in a cross-country study carried out among women in several fragile and conflict-affected (FCA) countries by Kabir and Klugman (2019), it was revealed that the adoption of DFSs, particularly through mobile devices, among women contributes to lowering gender gap in financial inclusion. Consequently, they present the mobile phone as an important tool for promoting financial inclusion, particularly among women with reduced mobility and those living in precarious environments.

While examining the effects of digital financial services on the access to and usage of financial services by women as well as on the general quality of financial services, the AFI (2020) in a study stated that DFSs can be employed to engender not just national financial inclusion but global financial inclusion. The study argued that since most DFSs (physical or real time) are transacted through a mobile interface, they enhance easy access to formal financial services at the customer's convenience, thereby making transaction costs affordable, reducing transportation costs and saving lost time for domestic responsibilities. In another study on the economic impact of gender equality, McKinsey Global Institute (2015) showed that closing the gender gap in digital financial inclusion, legal protection, education level and voluntary care work could help foster developmental strides. In particular, the study found that achieving gender equality could boost global annual GDP to $28 trillion by 2025.

Also, in a study on fostering women's digital financial inclusion, Imboden (2018) revealed that the constraints to women's financial inclusion include inadequate financial capability; financial illiteracy; low adaptation of financial products, services and delivery platforms to include women; women's lack of trust in formal financial institutions; collateral challenges; limited mobility; lack of formal mode of identification; lack of gender-disaggregated data; and cultural and legal limitations due to social norms. However, the study has shown that many of these challenges can be overcome by embracing digital finance. For example, digital financial services can address the lack of formal identification through the introduction of an advanced digital identification system; tackle collateral challenges by replacing dependence on conventional immoveable collateral with moveable collateral; enhance mobility by switching travel to long-distance bank outlets with financial services accessible through mobile phones; employ data from trade and utilities creditor to affirm the credit history of customers; and provide gender-specific innovative digital financial products and services.

Similarly, Global Partnership for Financial Inclusion [GPFI] (2020) claimed that the constraints to women's financial inclusion are aggravated by rules and norms that limit women's right to control assets, participate in the labour force, create and access credit for business expansion and become self-dependent. Further, compared to men, women are more likely to be poor and work in informal settings. However, the study found that despite the fact that many women around the world have been empowered by digital finance, the gender gap in digital financial inclusion still remains.

As a result, key policy options have been proposed to remove key barriers to women's assessment of financial services. Some of these key policy options include enhancing women's universal access to official identity systems and documents; promoting ownership of mobile phones and other devices through which women can access digital financial services; facilitating the deployment of infrastructure and procedure that enhance women's access to competitive government's digital transfers; strengthening the financial capability and digital skills of women through technology and change of behavioural perception; facilitating women's financial privacy and security to curb the risk associated with digital finance; abolishing legal discrimination against women's financial emancipation and removing irrelevant obstacles to opening financial accounts and women entrepreneurship; and offering appropriate incentives to female banking agents to increase financial institutions' female employees and to attract and retain women customers who prefer being served by their female counterparts.

In their study on the economic impact of digital finance, King and Petty (2021) find that through the uptake of digital financial services, notably the mobile money service known as M-Pesa, 185,000 Kenyan women have abandoned subsistence farming for business while 194,000 Kenyan households were lifted out of extreme poverty. Furthermore, the mobile money service has improved access to key credit facilities for business start-ups and has responded to emergencies. In a similar study conducted in South Africa, van Biljon et al. (2018) showed that the South African governments' digital transfers made through banking cards engendered greater financial inclusion of women, thereby increasing their decision-making power and participation in the economy.

Acknowledging the need for women's financial inclusion and its impact on their economic empowerment, Trivelli et al. (2018) bemoaned the persistent gender gap in access to formal financial services globally. Hence, the study proposed three key recommendations capable of removing the obstacles to women's financial inclusion in the G20 countries. They are supporting financial inclusion of women through the creation of digitally oriented financial solutions to remove mobility constraints and reduce financial transaction costs; developing measures to bridge the gender divide by encouraging the inclusion of the financially excluded and unbanked, especially women; and strengthening women-led businesses through their access to and usage of financial services.

The foregoing suggests a broad consensus on the positive impact that digital financial inclusion has on women's economic empowerment, labour market participation and, ultimately, global economic development. However, while the topic is critical, there is little research on the state of women's digital financial inclusion in Nigeria. As such, this chapter fills this gap by examining the gender dimension of access to and use of digital financial services and products in Nigeria.

12.3 Digital Financial Inclusion (DFI) in Nigeria: Stylized Facts

This section presents trends in the uptake of digital financial services in Nigeria to provide a gender perspective on digital financial inclusion. The Global Financial Inclusion Database presents a summary of financial inclusion indicators for all adults, based on specific demographic characteristics such as age, gender, employment status, income, education and locality. The database comprises over 850 national indicators of financial inclusion and covers over 140 countries. Furthermore, the first edition of the database was published in 2011, and updates to the data have been made every 3 years since then.

Based on an extract from the Global Financial Inclusion Index database, Table 12.1 shows the distribution of the access to various digital financial channels and services across genders in Nigeria. It shows that male ownership of financial institution accounts increased by about 21% between 2011 and 2014, but declined by 3% between 2014 and 2017. The narrative is not different for female ownership of financial institution accounts as an increase of about 8% was recorded between 2011 and 2014, with a decline of about 7% between 2014 and 2017. In general, ownership of financial institution accounts increased between 2011 and 2014, but decreased in 2017, with evident gender gaps. The general reduction in financial account ownership across genders may be partly attributed to the mandatory bank verification number (BVN) registration, which commenced in 2014 and might have resulted in the closure of unverified bank accounts.

In addition, although the percentage of people who paid bills on the Internet is generally low by gender, the percentage of men in this category is higher than that of women. Similarly, the percentage of people who bought online is also low for both men and women. This trend may be partly due to the low Internet penetration of the country which has deprived a large percentage of the populace of access to the Internet (Odunsanya and Adetutu 2020; Umezuruike et al. 2015). It may also be attributed to the low digital and financial literacy of the economy, which precludes most people from using the available Internet to purchase goods and services online (Olanrewaju and Afolabi 2022). In the same vein, the percentage of the population who made purchases on the Internet is also extremely low by gender for the review period. There was an increase of 4% and 1% in the percentage of the

Table 12.1 Access to digital financial services by gender in Nigeria (extracted)

	2011		2014		2017	
Digital financial inclusion indicators	Male	Female	Male	Female	Male	Female
Ownership of financial institution account	33%	26%	54%	34%	51%	27%
Paid bills using the internet	n.a	n.a	n.a	n.a	6%	2%
Made purchase online using the internet	n.a	n.a	5%	2%	9%	3%
Owned debit card	19%	18%	45%	25%	40%	23%
Owned credit card	1%	1%	4%	2%	3%	2%
Accessed account using the internet or mobile phone	n.a	n.a	n.a	n.a	10%	5%
Owned a mobile money account	n.a	n.a	3%	2%	7%	4%
Made digital payments	n.a	n.a	38%	20%	31%	17%
Received digital payments	n.a	n.a	34%	20%	29%	16%

Source: Extracted and Adapted from Global Financial Inclusion (GFI) Index (Demirgüç-Kunt et al. 2018)
Note: *n.a* denotes data not available

male and female population who used the Internet to make a purchase online between 2014 and 2017; nonetheless, the gender gap in this distribution still exists with the male gender having an upper hand.

Furthermore, there was a 1% gender gap in the percentage of men and women in possession of debit cards in 2011. However, by 2014, the gender gap had become more widened as 45% of the male population compared to 25% of the female population owned a debit card, indicating a gender difference of 15%. The percentage of men and women with debit cards declined by 5% and 2%, respectively, between 2014 and 2017. However, the gender gap widened by 2% during this period. This suggests that more male than female have access to debit cards in the country, and this may be partly attributed to the bottlenecks associated with obtaining debit cards in financial institutions at the time.

Most financial institutions require physical presence to process and obtain debit cards. This can be a daunting task for women because of their literacy rates, domestic responsibilities and certain legal and cultural barriers. As with debit card ownership, credit card ownership tends to be low by gender. The reason for this trend may not be far-fetched, since most transactions in the country do not require credit cards. In other words, the prevalence of credit card use is generally low in Nigeria, which explains its low rate for both men and women. However, irrespective of this fact, the table shows gender differences in credit card ownership for the whole review period.

Further, the table shows that in 2017, the percentage of the male population who accessed their account via the Internet or mobile phone dominated the female population. Specifically, the male population that accessed its accounts through the Internet or mobile phone doubled the female population at that time. On the other hand, a wide gender gap was also observed with respect to the proportion of the male and female population who owned a mobile financial account with the gender gap increasing from 1% to 3% between 2014 and 2017. Again, this low trend may be reflective of the low Internet penetration status of the country. Even though gender-specific data on Internet penetration in the country is not available, the World Bank (2020) database revealed that only 28% of the total population used the Internet in 2017, while mobile cellular subscriptions dropped by about 6.1% between 2016 and 2017. More so, according to the GSMA (2018) report, more men than women are likely to be in possession of a mobile phone in low- and middle-income economies including Nigeria, with cost of mobile phones, illiteracy, lack of digital skills, cost of credit and network coverage being the key limiting factors in this regard. However, even though the Internet penetration rate has since increased between 2017 and 2020, the growth rate is just approximately 29% indicating that the Internet penetration rate is still at a low ebb in the country.

At the same time, there is a large gap in the percentage of men and women who have made and received payments digitally. Interestingly, the proportion of the population who made digital payments slightly exceeds the digital payment recipients across gender, indicating that some of the financial transactions are done with foreign companies or organizations. This also shows the level of trade interaction of Nigerians with the rest of the world. In addition, the gender gap for digital payments was 18% in 2014 and decreased by 4% in 2017. Similarly, in 2014, the gender gap for payments received digitally was 14%, down 1% in 2017. The decline in digital transactions in Nigeria suggests the need to deploy strategies to improve digital infrastructures and create awareness on the economic benefits of using digital platforms for financial transactions.

Overall, the foregoing suggests the prevalence of the gender gap malady in digital financial inclusion in Nigeria as almost every digital financial services and digital finance channel available in the country has an associated gender gap. Since achieving universal digital financial inclusion is key to sustainable development, it becomes imperative that strategic measures be implemented to close the gender gap in Nigeria's digital financial inclusion. This would improve women's economic empowerment, advance gender equality and accelerate the pace of achievement of the SDGs in the country.

12.4 Bridging the Gender Gap in Digital Financial Inclusion in Nigeria: Lessons from International Experience and Policy Priorities

Success stories abound about different countries' strategies for achieving digital financial inclusion. These success stories, from which lessons are drawn for Nigeria, are reviewed in this section. Achieving gender parity in the access to and usage of digital financial services and products in Nigeria entails eliminating all barriers to women's digital financial inclusion through the collaborative efforts of the government, financial services industry, large businesses, and external donors and development partners (GPFI 2015). This is because the myriad of barriers women confront in their access to digital financial services may deter the potential contribution of digital finance to the SDGs and further expand the gender gap in financial inclusion (Molinier 2019).

As mentioned earlier, although the government has made considerable efforts in terms of enactment of policies geared towards reducing the financially excluded population, unfortunately, these policy targets, though differentiated by gender, have not yielded significant positive outcomes. Therefore, the government needs to do more to effectively address the financial gap between men and women in Nigeria. In addition to gender-specific policy development, the government should be consistent and committed to implementing policies to achieve desired objectives. Tanzania has made considerable progress in closing the gender gap in financial inclusion through the introduction of its 2013 National Financial Inclusion Framework, which provides a better financial environment for women. Tanzania is renowned globally for its favourable regulatory climate for financial inclusion which has expanded digital financial inclusion, through mobile money (AFI 2017). The country's financial inclusion strategy includes a regulatory framework for mobile money and financial infrastructure through which the financial inclusion of women was reinforced (AFI 2017).

Even though Nigeria's 2013 National Financial Inclusion Strategy (NFIS) targets were not gender specific, the government can take a cue from the Tanzanian government's initiative of adopting and implementing priority policies that support women digital financial inclusion, particularly through the use of mobile phones and wallets amongst the underserved women population in Nigeria. According to the Global Partnership for Financial Inclusion [GPFI] (2015), expanding women's digital financial inclusion will require the digitalization of the platforms by which government payments and transfers are made and received. This would open up opportunities for business expansion and innovation among low-income women. Further, the Nigerian government should be saddled with the responsibility of revamping the nation's digital financial infrastructural facilities and providing a conducive regulatory environment that permits full operationalization of mobile financial services while reforming discriminatory policies that hinder women's accessibility to and usage of digital financial services.

The financial services industry also has a key role to play in closing the gender gap in digital financial inclusion in Nigeria. According to AFI (2017), a financial service provider in Nigeria, Diamond Bank, in collaboration with Women's World Banking, developed an initiative through which financial transactions can be made via a mobile phone. This innovation addresses the needs of self-employed adults (market men and women) who desire to save regularly from their daily returns. While this initiative can be adjudged a good step in the right direction, it places a responsibility on Nigeria's financial services industry to consistently design and develop digital financial services and products that meet the unique needs and preferences of the female population. Evidence from successful countries in this regard shows that carefully designing products that focus on unique demands and social contexts are essential to enhancing the benefits of digital financial services to women. Thus, designing new products would entail identifying the target audience, in this case women, and conducting a needs assessment to recognize the most suitable design features for the targeted population (Dwitami 2021).

It is also imperative that the country's financial services industry be developed and encouraged by monetary authorities to promote innovation to improve women's digital financial inclusion. Specifically, innovative digital financial service providers should be incentivized regularly by the government and monetary authorities to encourage the development and promotion of novel DFSs and products that are attractive, affordable, smart and user-friendly for female folk. This is the case in Singapore, where the monetary authority adopts the regulatory sandbox mechanism in support of the financial services sector. The regulatory sandbox is a system that develops regulations for financial service businesses to support the rapid growth of industrial innovation. This mechanism helps financial technology firms, including those with products specifically for women, to conduct research on innovative financial services and products within a definite time and space, thus reducing the spate of missed opportunities for financial technology firms, who would otherwise desist from developing financial services and products for women that are potential loan defaulters (Alam et al. 2019). In addition, a robust and stable financial transaction network as well as sturdy customer support should be provided by financial service providers to assist in resolving complications that may arise while using the digital financial platforms and services (GPFI 2015).

Besides, novel technologies such as biometrics can be leveraged to address the low literacy and technical adoption rate among women while alternative credit scoring methods could be employed to grant women access to credit facilities and services (GPFI 2015). Nigeria can take a cue from countries across Asia and the Pacific region, who are leveraging digital technologies such as biometrics with its associated facial recognition and fingerprint technology to enhance their national identification strategy. The giant strides made in this region in this regard have helped in improving the delivery of financial services to women who otherwise lacked the basic requirements enforced by financial institutions (Salman and Nowacka 2020).

Further, barriers to the uptake of digital financial services and platforms stemming from poor Internet penetration and lack of mobile phones can be ameliorated by the introduction of low-tech innovative techniques by financial services providers. Through this means, a number of financially excluded women in the Asia and Pacific region who reside and work in rural areas with low Internet and mobile connectivity have been reached (Salman and Nowacka 2020).

Moreover, large firms can leverage their size and high level of financial transaction to enhance the uptake of digital financial services among females by disbursing remuneration via digital channels, encouraging electronic payments and creating a digital supply chain. In addition, large companies should be encouraged to provide services via the Internet or mobile devices to facilitate women's ownership of mobile digital devices (GPFI 2015).

In addition, external donors and development partners in Nigeria are called upon to support policies and regulatory frameworks that encourage women's digital financial inclusion. Countries like Nigeria should also receive technical assistance to build capacity to address the risks associated with women's use of digital financial innovations. There is also a need to organize capacity-building initiatives that pay attention to the skill and knowledge gaps as well as attitude of women towards the adoption of digital financial services. This would go a long way in accelerating the pace of digital financial inclusion among women in Nigeria.

In summary, the following policy priorities should assist the Nigerian government and other stakeholders in bridging the gender gap in financial inclusion and empowering women:

1. Given that mobile phones provide a rapid expansion of digital financial inclusion, and many who lack access to formal finance have

a phone, stakeholders should facilitate women's universal ownership of mobile phones.
2. Given that the keys to building an inclusive digital payments ecosystem are competition and interoperability, stakeholders should promote efforts to implement infrastructure and protocols for digital government payments to women that are competitive and interoperable with private sector payment systems.
3. Given that there are both potential benefits for women from accessing and using digital financial services and certain risks that need to be managed, stakeholders should leverage technological and behavioural knowledge to build women's digital skills and financial capabilities.

12.5 Conclusion

Gender parity in digital financial inclusion is key to the achievement of the United Nations Sustainable Development Goals. But the wave of gender gap in digital financial inclusion around the world is alarming and damaging to development imperatives. Therefore, this chapter examined the gender dimensions of digital financial inclusion in Nigeria with the aim of evaluating the role of digital finance in bridging the pervasive gender divide in financial inclusion in Nigeria. The results demonstrated that virtually all digital financial services and digital finance channels available in the country have an associated gender gap. Specifically, the gender gap was more evident in the percentage of the male and female population who owned a debit card, owned a financial institution account and made or received digital payments in the period under review. This trend has the potential to undermine the achievement of the Sustainable Development Goals in the country by 2030. Thus, taking cues from countries who have been able to make significant headway in bridging the gender gap in digital financial inclusion, the chapter provides key policy recommendations which require the collaborative efforts of the Nigerian government, financial services industry, large firms, external donors and development partners, as a more meaningful financial inclusion has proved to be the key to unlocking the potentials for promoting more sustainable and equitable development.

References

Alam N, Gupta L, Zameni A (2019) Fintech and Islamic finance. Palgrave Macmillan, Cham

Alliance for Financial Inclusion (2017) Bridging the gender gap: promoting women's financial inclusion. https://www.afi-global.org/wp-content/uploads/publications/2017-11/AFI2017_Gender_full_AW_ISBN_digital.pdf

Alliance for Financial Inclusion (2020) Lessons on enhancing women's financial inclusion using Digital Financial Services (DFS). https://www.afi-global.org/sites/default/files/publications/2020-05/AFI_WFI_DFS_SR_AW_digital.pdf

Aluko Y (2018/2019) Dimensions of exclusion and inclusion of women in Nigeria. Annals of the Social Academy Nos 22, Jan.–Dec., pp 34–59

Demirgüç-Kunt A, Klapper L, Singer D, Ansar S, Hess J (2018) The global findex database 2017: measuring financial inclusion and the Fintech revolution. World Bank, Washington, DC

Dwitami AP (2021) Advancing inclusive digital financial services to empower women in Indonesia. https://www.povertyactionlab.org/blog/10-1-21/ifii-blog-series-advancing-inclusive-digital-financial-services-empower-women

Ferrata L (2019) Digital financial inclusion—an engine for "leaving no one behind". Public Sect Econ 43:445–458

Gammage S, Kes A, Winograd L, Sultana N, Hiller S, Bourgault S (2017) Gender and digital financial inclusion: What do we know and what do we need to know? International Center for Research on Women (ICRW)

Ghoh S (2022) Political empowerment of women and financial inclusion: is there a link? Soc Sci Humanit Open 5(1):2022. https://doi.org/10.1016/j.ssaho.2022.100267

Global Partnership for Financial Inclusion (2015) Digital financial solutions to advance women's economic participation. http://www.g20.utoronto.ca/2015/Digital-Financial-Solution-to-Advance-Women....pdf

Global Partnership for Financial Inclusion (2016) Global Standard-Setting Bodies and Financial Inclusion: The Evolving Landscape. https://www.gpfi.org/sites/gpfi/files/documents/GPFI_WhitePaper_Mar2016.pdf

Global Partnership for Financial Inclusion (2020) Financial inclusion action plan. October 2020

GSMA (2018) The mobile gender report, 2018. https://www.gsma.com/mobilefordevelopment/wp-content/uploads/2018/04/GSMA_The_Mobile_Gender_Gap_Report_2018_32pp_WEBv7.pdf

Helms B (2006) Access for all: building inclusive financial systems. The World Bank. https://doi.org/10.1596/978-0-8213-6360-7

Imboden K (2018) Advancing women's digital financial inclusion. https://res.torontocentre.org/guidedocs/Advancing%20Womens%20Digital%20Financial%20Inclusion%20FINAL.pdf

Kabir R, Klugman J (2019) Women's financial inclusion in a digital world: how mobile phones can reduce gender gaps. Georgetown Institute for Women, Peace and Security (GIWPS), Washington, DC

King B, Petty R (2021) Financial inclusion through digital inclusion. https://www.ifac.org/knowledge-gateway/preparing-future-ready-professionals/discussion/financial-inclusion-through-digital-inclusion

Lauer K, Lyman T (2015) Digital financial inclusion: implications for customers, regulators, supervisors, and standard-setting bodies. https://www.cgap.org/sites/default/files/researches/documents/Brief-Digital-Financial-Inclusion-Feb-2015.pdf

Mckinsey Global Institute (2015) The power of parity: how advancing women's equality can add $12 trillion to global growth. https://www.mckinsey.com/~/media/mckinsey/featured%20insights/employment%20and%20growth/how%20advancing%20womens%20equality%20can%20add%2012%20trillion%20to%20global%20growth/mgi%20power%20of%20parity_full%20report_september%202015.pdf

Molinier H (2019) Leveraging digital finance for gender equality and women's empowerment. https://www.unwomen.org/sites/default/files/Headquarters/Attachments/Sections/Library/Publications/2019/Leveraging-digital-finance-for-gender-equality-and-womens-empowerment-en.pdf

Naumenkova S, Mishchenko S, Dorofeiev D (2019) Digital financial inclusion: evidence from Ukraine. Invest Manag Financ Innov 16:194–205

Odusanya K, Adetutu M (2020) Exploring the determinants of internet usage in Nigeria: a micro-spatial approach. In: Hattingh M, Matthee M, Smuts H, Pappas I, Dwivedi Y, Mäntymäki M (eds) Responsible design, implementation and use of information and communication technology, pp 307–318. https://doi.org/10.1007/978-3-030-45002-1_26

Olanrewaju B, Afolabi J (2022) Digitizing education in Nigeria: lessons from COVID-19. Int J Technol Enhanc Learn 14:402–419

Omar MA, Inaba K (2020) Does financial inclusion reduce poverty and income inequality in developing countries? A panel data analysis. Econ Struct 9:1–25

Ozili P (2022) Digital financial inclusion. https://www.researchgate.net/publication/358221520_Digital_Financial_Inclusion

Riley P, Romorini S, Golub E, Stokes M (2020) Digital financial services in the MENA region. Sustaining Health Outcomes through the Private Sector Plus Project, Abt Associates Inc, Rockville, MD

Salman A, Nowacka K (2020) Innovative financial products and services for women in Asia and the Pacific. ADB Sustainable Development Working Paper Series 67:1–88

Trivelli C, Villanueva C, Marincioni M, Pels J, Sachetti F, Robino C; Walbey H, Martinez L, Magnelli M (2018) Financial inclusion for women: a way forward. https://t20argentina.org/wp-content/uploads/2018/07/Financial-Inclusion-for-Women-Final.pdf

Umezuruike C, Awodele O, Izang A (2015) Broadband internet penetration in Nigeria: a review. Int J Res Stud Comput Sci Eng 2:2349–4859

van Biljon C, von Fintel D, Pasha A (2018) Bargaining to work: the effect of female autonomy on female labour supply. Working papers 04/2018, Stellenbosch University, Department of Economics

World Development Indicator. World development indicator, Nigeria. 2020. https://data.worldbank.org/indicator/IT.NET.USER.ZS?locations=NG

Index

T. A. Ojo (ed.), *Women and Finance in Africa*, Sustainable Development Goals Series, https://doi.org/10.1007/978-3-031-53337-2

SPRINGER NATURE

GPSR Compliance

The European Union's (EU) General Product Safety Regulation (GPSR) is a set of rules that requires consumer products to be safe and our obligations to ensure this.

If you have any concerns about our products, you can contact us on ProductSafety@springernature.com

In case Publisher is established outside the EU, the EU authorized representative is:

Springer Nature Customer Service Center GmbH
Europaplatz 3
69115 Heidelberg, Germany

The manufacturer's authorised representative in the EU is Springer Nature Customer Service Centre GmbH, Europaplatz 3, 69115 Heidelberg, Germany. If you have any concerns regarding our products, please contact ProductSafety@springernature.com

Printed and bound by CPI Group (UK) Ltd, Croydon, CR0 4YY
01/10/2026
02225919-0020